Planet Word

Planet Word

J. P. DAVIDSON

*with a foreword
by Stephen Fry*

MICHAEL JOSEPH
an imprint of
PENGUIN BOOKS

MICHAEL JOSEPH

Published by the Penguin Group
Penguin Books Ltd, 80 Strand, London wc2r orl, England
Penguin Group (USA) Inc., 375 Hudson Street, New York, New York 10014, USA
Penguin Group (Canada), 90 Eglinton Avenue East, Suite 700, Toronto, Ontario, Canada m4p 2y3
(a division of Pearson Penguin Canada Inc.)
Penguin Ireland, 25 St Stephen's Green, Dublin 2, Ireland (a division of Penguin Books Ltd)
Penguin Group (Australia), 250 Camberwell Road,
Camberwell, Victoria 3124, Australia (a division of Pearson Australia Group Pty Ltd)
Penguin Books India Pvt Ltd, 11 Community Centre,
Panchsheel Park, New Delhi – 110 017, India
Penguin Group (NZ), 67 Apollo Drive, Rosedale, Auckland 0632, New Zealand
(a division of Pearson New Zealand Ltd)
Penguin Books (South Africa) (Pty) Ltd, 24 Sturdee Avenue,
Rosebank, Johannesburg 2196, South Africa

Penguin Books Ltd, Registered Offices: 80 Strand, London wc2r orl, England

www.penguin.com

First published 2011

1

Printed in Germany by Firmengruppe Appl, aprinta druck Germany

A CIP catalogue record for this book is available from the British Library

isbn: 978–0–718– 15774–6

www.greenpenguin.co.uk

Contents

List of Illustrations

Chapter 1

Tower of Babel (Mary Evans Picture Library)
Ambam (J. P. Davidson)
Toby the Sapient Pig (Mary Evans Picture Library)
Hans the Counting Horse (Mary Evans Picture Library)
Navy-trained dolphin (Photo by Evening Standard/Getty Images)
Nim Chimpsky (Susan Kuklin/Science Photo Library)
Brain diagram (National Institute on Deafness and Other
  Communication Disorders, www.nidcd.nih.gov)
Noam Chomsky (Photo by Ulf Andersen/Getty Images)
FoxP2 gene (Ramon Andrade 3Dciencia/Science Photo Library)
Steven Pinker (Photograph by Francesco Guidicini, Camera Press
  London)
This is a wug (Copyright Jean Berko Gleason)
Akbar the Great (Private Collection/ Peter Newark Pictures/ The
  Bridgeman Art Library)
Wild Boy of Averyon (Science Photo Library)
GB sign language (© 2011 Cath Smith – from the LET'S SIGN
  Series – www.deafbooks.co.uk)
Abbé Charles-Michel de l'Epée (Mary Evans Picture Library/Alamy)
National Theatre of the Deaf, Connecticut (J. P. Davidson)
Wilhelm Grimm (The Print Collector/Heritage-Images/Scala,
  Florence)
Jacob Grimm (The Print Collector/Heritage-Images/Scala, Florence)
Dornröschen (Mary Evans Picture Library)
Brüder Grimm Kinder-Märchen (Mary Evans Picture Library)
Grimms' Fairy Tales (TopFoto)
Kinder und Hausmärchen (Mary Evans Picture Library)
United Nations headquarters (Emmanuel Dunand/AFP/Getty
  Images)

Chapter 2

Turkana Woman with beads (J. P. Davidson)
English and Swahili newspapers (Impact Photos/Alamy)
Turkana teacher and children (J. P. Davidson)
Turkana elder and children (J. P. Davidson)
Akha houses (J. P. Davidson)
Aju (J. P. Davidson)
Akha school (J. P. Davidson)
Boy and girl at Carera (Getty Images)
Funeral procession (Photo by Illustrated London News/Hulton
   Archive/Getty Images)
Jonathan Swift (Getty Images)
Lady Gregory (George C. Beresford/Getty Images)
Autograph Tree (Time & Life Pictures/Getty Images)
Abbey Theatre (Hutton Archive/Getty Images)
Ros na Rún (J. P. Davidson)
Irish posters (Time Life Pictures/Getty Images)
Beowulf (Scala Florence/Heritage Images)
'Speak French, Be Clean'
Welsh Not (St Fagans: National History Museum)
Irish tally sticks (Science Museum/SSPL)
Frédéric Mistral (© Photos 12/Alamy)
Académie Française (Mary Evans Picture Library)
Book of Samuel (The Jewish Museum/Art Resource/Scala, Florence)
Hebrew Language Council (Scala Florence/Heritage Images)
Ticket to 1908 Yiddish Language Conference
Woody Allen and Billy Crystal (The Moviestore Collection Ltd)
Ludovic Zamenhof (Mary Evans/AISA Media)
Reĝo Lear (Image courtesy of Universala Esperanto-Asocio)
Stephen Fry and Dr Mark Okrand (J. P. Davidson)
Star Trek III (The Moviestore Collection Ltd)
Wilfred Pickles portrait (Photo by Bert Hardy/Getty Images)
Wilfred Pickles cartoon (© IPC+ Syndication)
Queen's Christmas broadcast, 1952 (Getty Images)
Ant and Dec (Allstar Picture Library/Alamy)

Chapter 3

George Carlin (Getty Images)
Hunters and bear (Mary Evans Picture Library)
Deaf Tourette's (Alistair Richardson)
Mary Whitehouse (Getty Images)
Yes, Minister (Copyright BBC)
The Thick of It (Copyright BBC)
Edward de Vere (Getty Images)
Marie Lloyd (Mary Evans Picture Library)
Max Miller (TopFoto)
The BBC's 'Little Green Book' (BBC Picture Library)
Etiquette in Everyday Life (Amoret Tanner/Alamy)
Legal jargon cartoon (www.CartoonStock.com)
1984 (The Moviestore Collection Ltd)
Guy Gibson and Nigger (Photo by Associated Newspapers/Daily
    Mail/Rex Features)
Love Thy Neighbour (FremantleMedia Ltd/Rex Features)
Francis Grose (The Print Collector / Heritage-Images)
Costermongers (Mary Evans Picture Library/David Lewis Hodgson)
Only Fools and Horses (The Moviestore Collection Ltd)
Bow Bells (Mary Evans Picture Library)
Tod Sloan (Mary Evans Picture Library)
Round the Horne (Copyright BBC)
Crocodile Dundee (The Moviestore Collection Ltd)
The Adventures of Barry McKenzie (Alamy)
I heart NY (Jon Arnold Images Ltd/Alamy)
Homer Simpson (20th Century Fox/The Kobal Collection/Groening,
    Matt)
Ali G (Tony Kyriacou/Rex Features)
Hip-hop (© Laurence Watson/PYMCA)

Chapter 4

Cuneiform tablet (© The Trustees of the British Museum)
Clay tablet (© The Trustees of the British Museum)
Georg Friedrich Grotefend (akg-images)
Jean-François Champollion (IBL BildbyrÕ/Heritage-Images/TopFoto)
Rosetta Stone (akg-images/Erich Lessing)
Linear B (Lawrence Lo, Ancient Scripts of the World, http://www.
  ancientscripts.com)
Michael Ventris (Getty Images)
Phoenician alphabet (Alamy)
Dead Sea Scrolls (Robert Harding)
Ad from *Wall Street Journal*
Yigael Yadin (Getty Images)
Stephen Fry with scroll (J. P. Davidson)
Chinese Oracle bones (Getty Images)
Zhou Youdong (J. P. Davidson)
Diamond Sutra (© The Trustees of the British Museum)
Gutenberg Bible (© The British Library Board, shelfmark C.9.d.4)
Caxton *Canterbury Tales* (Scala Archives)
Café Procope (Getty Images)
Encyclopédie (TopFoto)
Samuel Johnson (Mary Evans Picture Library)
James Murray (By Permission of the Secretary to the Delegates of
  Oxford University Press)
James Murray and his assistants (By Permission of the Secretary to the
  Delegates of Oxford University Press)
Library of Alexandria (Alamy)
The Bodleian (akg-images)
Andrew Carnegie (National Portrait Gallery, London)
The Pickwick Papers (IAM/akg/World History Archive)
Dick Turpin (Mary Evans Picture Library/Edwin Mullan Collection)
The Union Jack (Getty Images)
Detective Magazine, The Grange Collection, New York/TopFoto
iPad artwork (Matthew Young)
Mr Yuk (Used by permission from the Pittsburgh Poison Center,
  University of Pittsburgh Medical Center, Pittsburgh, PA, USA)
Poison cartoons (Jon Lomberg)

Chapter 5

Turkana warriors (J. P. Davidson)

Aborigine songlines (Mary Evans Picture Library)

Homer's *Odyssey* (Mary Evans/Rue des Archives/Tallandier)

Homer (Mary Evans Picture Library)

Achilles (Mary Evans Picture Library)

Bonnie and Clyde (Warner Bros/ RGA)

William Goldman (Getty Images)

Butch Cassidy and the Sundance Kid (20th Century Fox/The Kobal Collection)

James Joyce (Mary Evans/AISA Media)

W. B. Yeats (Photo by Spicer-Simson/Hulton Archive/Getty Images)

The Globe Theatre (Süddeutsche Zeitung/Mary Evans Picture Library)

Laurence Olivier (Mary Evans Picture Library)

Stephen Fry and Simon Russell Beale (Sandi Friend)

David Tennant (Donald Cooper/Rex Features)

Mark Rylance (Richard Mildenhall)

Bob Dylan (Mary Evans/Interfoto/Lisa Law)

Paul Simon and Art Garfunkel (Photo by Pictorial Parade/Getty Images)

Crowd singing (Action Plus Sports Images / Alamy)

Wedgwood ad (By Courtesy of The Wedgwood Museum)

Lemon car ad image (Courtesy of The Advertising Archive)

'Go to work on an egg', Marmite, KitKat, Heinz ads (Image Courtesy of The Advertising Archives)

David Ogilvy (© Bettmann/Corbis)

Mad Men's Don Draper (Amc/The Kobal Collection)

The Hathaway Man (Image Courtesy of The Advertising Archives)

Jesse Jackson (Canadian Press/Rex Features)

Barack Obama (Paul J. Richards/Afp/Getty Images)

Chief Joseph (Mary Evans Picture Library)

Winston Churchill (Mary Evans Picture Library)

Went the Day Well? (Ealing Studios/Ronald Grant)

1984 (Columbia/The Kobal Collection)

Adolf Hitler (Photo by Keystone/Getty Images)

German Propraganda (Mary Evans Picture Library/Explorer Archives)

Chairman Mao (Photo by Universal History Archive/Getty Images)

'Stalin Leads Us to Victory' (RIA Novosti / TopFoto)

'La parole humaine est comme un chaudron fêlé où nous battons des mélodies à faire danser les ours, quand on voudrait attendrir les étoiles.'

('Human language is like a cracked kettle on which we beat out tunes to make bears dance when what we really want is to move the stars.')

Gustave Flaubert, Madame Bovary (1857)

Foreword

by Stephen Fry

Words are all we have. Certainly, reader, words are all *we* have, you and I, as you sit with this book or reading device in front of you and I sit and tap at my keyboard. You have no idea where I am as I do this, and I have no idea who, where or what you are as you continue to read. We are connected by a filament of language that stretches from somewhere inside my brain to somewhere inside yours. There are so many cognitive and cerebral processes involved simply in the act of my writing and your reading these words that not all the massed ranks of biology, genetics, linguistics, neurology, computational science or philosophy can properly describe, let alone understand or explain, how it all works.

Yet language, as we are all aware, is a human birthright. It is as free and available to us all as the ability to walk. The impairment, trauma, abuse or psychological impediment must be very severe indeed for any child in the world not to be able acquire their language with no more effort (and usually less pain) than they acquire their teeth.

Language has delighted, enthralled and enraptured me since I can remember. I sometimes imagine that I have been granted by nature a greater awareness and higher sense of language as compensation for my appalling deficiencies in music, mathematics and all things athletic. Musical athletes who speak and write well are a sore aggravation to me, as you might imagine.

From the earliest age I played with language as others played with toy cars and guns. Constant repeating, altering, distorting, rhyming, punning, inventing and vamping with words was as natural to me (and doubtless as irritating to others) as sniffing lampposts is to dogs or air-guitaring and football was to normal boys.

Language was my way of both getting me into trouble and getting me out of it. To learn a new word was to make a new friend, twee as that may sound. I was a verbal dandy, unquestionably, but the diffusion of pleasure that spread through me as I learned and thereby 'possessed' new words was real and impossibly thrilling. I can still see in my mind's eye the actual positioning of certain words on the pages of the *Concise Oxford Dictionary* that was my constant companion from the age of eight to eighteen. Words like *prolix, strobile, banausic* and *pleonasm.* Intolerable show-off that I was, I peacocked them at every opportunity, but they *mattered* to me. This ownership of new words, coupled with the tracing back of their lineage, 'etymology' as I learned to call it, gripped me entirely.

'Did you know,' I would ask a bored friend, 'that the word "sycophant" originally meant someone who showed figs?' and the bored friend would say 'Wow' in that way that people do when what they really mean is 'I wish you would fall into a coma for ever'.

It struck me from an early age, and the belief has never left me, that language is not *celebrated* enough. As a study, linguistics has flourished (despite being bogged down for so long in the rather arid schism of Chomskyites versus Whorfians, of which more later), splitting into psycho-, neuro- and socio-crossbreeds that have a presence at most serious universities; crosswords and language games like Boggle and Scrabble thrive more than ever; discussions about 'correct' use and the 'dumbing down' of language still take up a disproportionate amount of readers' letters space in the newspapers – all this is certainly true, but how rarely do people *play* or *perform* with their nature-given power of utterance in the way they might play or perform with music by dancing in clubs and at festivals, by walking everywhere with tunes in their ears and by whistling and humming as they shower? Even people who can't draw can doodle. But how often do we doodle with language? I do it all the time. If you were to plant a recording device in my living quarters, you would think me insane as I verbally nonsensed my way around the flat. Perhaps my plea for playfulness with language is a plea for me to be thought more normal, but it seems to me that language should be the last of all human attributes to be taken for granted.

What I am saying, I suppose, is that with this astonishing resource readily available and costing no more than breath or finger-tapping to produce, why is it that most people are so dull and unadventurous in their use of it? Why don't they delight themselves with new yokings and phrases, new rhythms and coinings, new pronunciations and abbreviations?

Well, one answer is that collectively they do. All the time. Especially when young. Most juvenescent sodalities (two others words I learned early and overused embarrassingly), most social groupings of young people, have their own private language, catchphrases and nicknames for people and processes. But language (especially English perhaps) presents a problem. Embarrassment, shame, a sense of inferiority, unfashionable regionality, gender, sexuality, age, education – all these dreadful bugaboos come into play whenever we exchange language outside our, for want of a better phrase, 'peer group' and we lose confidence in the creative side of our linguistic selves for fear of the negative judgements and snobbish contempt of the mainstream, just as we might one day lose our piercings and the coloured streaks in our hair.

All that bright individual verbal clothing is put away for the workplace and dull, pretentious verbal suits are worn in their place. Never was the word 'suit' less . . . well . . . suitable. The memos, meetings and conferences of the workplace are couched in agglomerations of phrases as soulless, bloodless, styleless and depressing as the grey carpets, strip lighting and hessian partitions that constitute their physical environment. Sick-building syndrome is now well understood, sick language syndrome perhaps less so.

But this is to talk about language within language within *a* language. When I told friends that I was off to make a series of films about language for the BBC, the most common response was 'Which languages?' and I became used to having to explain that I meant programmes looking at Language with a capital L, which must of course include individual languages, but would hope to look at the nature of the phenomenon, the achievement, the *gift* of language itself. Where it comes from, how it split into the 6,000 world tongues we now have and why some of these are disappearing by the hundreds every year; how language is acquired in each human individual; how it is used for persuasion, tyranny, solace, art and commerce; how or if the nature of one tongue influences, defines or circumscribes the actual thought of its user; how we respond to its transgressive deployment in blasphemy, obscenity and other offensive usages . . . so many questions, so many areas of interest, such an endlessly fascinating and elusive subject.

This book reflects the major quests that I and J. P. Davidson, its author, who was also the producer of the series and directed three of the programmes, set out on. Over the course of six months or so, we and Helen Williamson, who directed the other two episodes, travelled the globe in search of all kinds of answers. None of us is able to say, any more than the most gifted and informed linguist can, whether English will be the dominant world tongue that it is today in a hundred years' time. We cannot predict the future trajectories of Mandarin, Arabic or Spanish. Warfare, famine, technology, trade and natural disasters have all played and will continue to play their decisive parts in linguistic dissemination and desiccation and the individual has yet to be born who can successfully predict the momentous upheavals in human affairs that drive people and their languages in new directions. Neither can we predict the new influences that will be brought to bear within individual languages. Who knew, just ten years ago, of OMG, lulz and retweeting? Nor can we know what words will be offensive to generations yet unborn. I can write the word 'fuck' now in the more or less certain knowledge that only an odd few (and I do mean *odd*) will be offended by the word. I cannot, however, write the word that begins with N and rhymes with 'Tigger' without blushing to my roots and fearing for my reputation. Some call this 'political correctness'. They are, in my estimation, deluded: their sense of language seems to be defined by the asininities of the worst of the bourgeois tabloids. Verbal taboos are far too interesting and complex to be fobbed off as fashionable liberal courtesies or even as simply 'good manners'.

If we can't tell the world anything new about language, why should we make the effort to produce five hours' worth of television and a big handsome book to accompany it? Because the *questions* that language raises are so much more eternally fascinating, revealing and beguiling than any theoretical answers and because, narratively speaking, as J. P. Davidson demonstrates in the following pages and as I hope we all demonstrated in the making of our programmes, *showing* is so much more interesting than *telling*.

Perhaps the biggest discovery I made, or at least the feeling I already had that was most heavily reinforced, was to do with language and identity. We may be what we eat, but we most certainly are what we say. Which is not to say I am taking sides on the schism I referred to earlier. All who know a little about linguistics will be aware of the Sapir–Whorf Hypothesis on the one hand and the Chomskian ideas of 'generative grammar' (the automatic innateness of language that is as programmed and predictable in its growth in a toddler as the arrival of hair is in the armpits of an adolescent) and the assertion that all languages, when it comes down to the essentials, are similar almost to the point of congruency. Against this is pitted the Whorfian notion that the very nature of the distinct language we speak determines, to a greater or lesser extent, the way we think or the way we see and interpret the world around us. The Chomskian view, as expressed so fluently, accessibly and convincingly by Steven Pinker, has held sway with academia for decades, but some research appears to have chipped away at the marmoreal smoothness of its surface lately. Wholly fascinating as this area of inquiry is, my sense of language as identity is more to do with the short time I spent among the Turkana in Kenya, the Akha in Thailand, the Basques of Spain and France and Irish-speakers in the Gaeltacht.

Our individual language may or may not limit or widen our thought according to its breadth of vocabulary, elasticity of structure or complexity of syntax, but it seems most certainly to *place* us in the world like no other property or quality we possess. In our limited and foolish way, we may think skin colour a greater determinant of identity, but an Ibo would feel no more in common with a Jamaican, I submit, than he would with me.

It so happened that I was in Kenya at the time of Barack Obama's election as president. I spoke to a member of the Luo tribe, from which Obama's father came, and asked if he was pleased that America should not only now have a black president, but one from his people. 'Very pleased of course,' came the reply, 'but you should consider that had Mr Obama been elected president of Kenya, he would have been our first *white* president.' Our confusion, inconsistency and insanity when it comes to labelling people as black when they are half or even three-quarters white, may one day, it is to be hoped, resolve itself into sense. True identity, aside from the very personal individual qualities, the DNA and parentage that separate all humans each from the other, resides in one cultural marker above all: language.

In the course of our travels I met in Beijing the most influential linguist who ever lived. One hundred and six years old, his first words to me were, 'You will have to forgive me, my English is a little rusty these days.' He modestly repudiated my claims for his place in world history, but I believe them nonetheless. He is the reason, incidentally, that we now write 'Beijing' and not 'Peking'. In London I underwent MRI trials that tried to locate the areas of my brain that were responsible for different types of conscious and automatic utterance. In Victoria, Australia, I attempted to get my befuddled mind around the absolute directional concepts built into the language of the Aboriginal people of Pormpuraaw. In Jerusalem I came as close to handling a fragment of the Dead Sea Scrolls as any human being can (only four in the world are allowed that actual privilege) and at the Max Planck Institute in Leipzig I had dung hurled at me by gorillas. I narrowly avoided smoking an opium pipe and eating curried dog on the borders of Thailand and Myanmar, and helped prepare a Basque-style, Michelin three-star lunch in San Sebastián. If you feel a tinge of envy, or something stronger, I cannot blame you. It was the gig of a lifetime and I am fully aware that it is no use my saying that it was very hard work, that the days were exhausting and the living conditions often atrocious. I am a lucky, lucky devil to have had such an opportunity.

I mentioned that language was, in my estimation, our clearest indicator of culture. Perhaps food comes a close second. The analogy holds at many levels. Just as one kind of cheap Western catering in the form of burgers, fried chicken and fizzy cola swamps towns and cities the world over, threatening natural indigenous cuisines, so one kind of English seems to be doing the same to minority languages. But, in a positive and countervailing manner, just as our bland English cuisine has been enriched, coloured and spiced by foreign influences from the world over, so too has our language. When families and individuals express their sense of who they are, it is as much through their mother's cooking as their mother tongue.

Whoever you are, whatever your provenance, however you came to be in the position you are now in, with this book or digital device in your hand, you can read and speak. What is more, the language you read and speak, while universally understood and given descriptive (but not prescriptive or proscriptive) grammatical rules and semantic definitions, is at one and the same time entirely your own and that of your clan, your tribe, your nation and your people. The way you speak is who you are and the tones of your voice and the tricks of your emailing and tweeting and letter-writing can be recognized unmistakably in the minds of those who know and love you.

I sometimes wonder if Alexander Pope should not have written that the proper study of mankind is *language*.

There was the television series, and now here in your hands is the book that celebrates language. Skip about or read it from first page to the last. We hope it will delight you and perhaps make you think afresh about the free, inexhaustible and delicious resource that lies somewhere in your brain and allows you to be who you are.

But next time you speak or write, do not try to work out what is going on socially, culturally, neurally, intellectually or physiologically. The effort is beyond us all and you might just explode. Instead . . . *celebrate*.

CHAPTER I

Origins

L et's start at the beginning, where all good stories should start. The trouble is, of course, we don't know when or how language started. All we know is, language evolved, just like we did, and continues to do so. But there are plenty of stories – myths of origin from the Bible, folklore and oral tribal traditions – which try to explain why we have so many languages. The Tower of Babel story in Genesis is probably the most famous. Versions of it appear throughout different cultures, but the one in Genesis has a precise function: to reconcile a creation myth with the existence of the extraordinary variety of languages, often living side by side. The priesthoods and shamans of many religions (not just the Abrahamic ones) had to create myths to explain how and why there are so many languages.

The story in Genesis goes that the people on earth did originally speak one language and live together. They built a city, and then a tower as a kind of communal rallying point, so high that it stretched up to heaven. God saw this as a dangerous sign of the ambitions and strength of men, so decided to break their unity by breaking their mutual understanding. He went down and 'confounded their language that they may not understand one another's speech and so the Lord scattered them abroad from thence upon the face of all the earth'.

An interesting origin myth is that of the Snohomish tribe, a Salishan-speaking people of the Pacific North-west of America. To start with, it's not so different from the creation act in Genesis. Their Sky

The Tower of Babel, after which language divided

Chief makes the world in pretty standard fashion: he gets some clay, rolls it into a big ball, covers it with soil, makes heavens and an underworld, connects all three with a world tree and then makes all the animals, including humans – man first, then woman out of man's tail. After that it gets creative. People start arguing whether the noise that flying ducks make comes from the beating of their wings or the wind blowing through their beaks. The chief calls a council to resolve the issue. The arguments become heated and bitter, and in the end they cannot agree, so the tribe splits up into wing-beaters, beak-blowers and agnostics. They scatter across the land, creating new communities. And this, so it is told, is the beginning of all the different tribes and tongues of people.

The profusion of languages amongst our species was, and continues to be, an extraordinary feat, putting new hunting tools and techniques in the shade. It is our defining achievement. But in a scientific sense how did we evolve to be the talking ape?

In the 50,000 to 100,000 years since we first started making sounds anything more convincing than ughs and grrs, many languages have come and gone. There are over 6,000 languages spoken on the planet today – some by more than a billion people, some by only a handful.

Languages evolved as rapidly as *homo sapiens* spread over the planet. It is the power of language which is crucial to man's survival: the grammars with their future tenses and conditional clauses that have empowered our species to think in terms of possibility, to hope and reach beyond the extinction of the individual. It has cemented tribal identity and allowed communal activities unlike those of any other species.

With language we build fictions, and an 'otherness' from our consciousness. And this, in a Darwinian sense, is of immense benefit: it is the essence of our creativity. It's also the glue that binds us together as social animals. Each language has its own way to articulate reality and dreams and so create poetry, myth, history and laws. Memory is held in language, and that defies time.

Language throws up many questions that generations of philosophers and scientists have sought to investigate. One of the trickiest is whether we can think without language, and there is no 'correct' answer. Nor is there one to the question of whether the language we speak affects the way we think. Does a German speaker have a more technical way of thinking, or a Frenchman a better understanding of love because they have written about it so much?

The other big debate is whether language is innate, part of our nature, our DNA, or a product of our environment, nurture. It's extraordinary that, within a matter of months, a mewling and cooing baby will begin to explore language with all its complexities of vocabulary, syntax, grammar, phonetics and metaphors. Remarkably not much later than it takes the average child to learn to walk, it will also have begun to use pronouns, and by the time the child is three to four years old it will not only comprehend stories and commands but will be able to construct relatively complex syntactically correct sentences with adjectives, prepositions and verb tenses. And what is truly astonishing is that the child won't even notice it's doing it.

What is even more amazing is that, if you plonk any young child down in a foreign culture and within a matter of months it will be speaking that language as fluently as any other person there. It might well forget it after a few years, especially if it's not used, but it will have acquired a language and in the process exercised probably the most complex bit of brain processing that we do.

So how on earth do we do it? There are plenty of theories thought up by legions of linguists, cognitive psychologists, neuroscientists, geneticists and evolutionary anthropologists, who believe they may hold some of the answers to the question of how it is that we humans can talk. There are those, like the linguist Noam Chomsky and his followers, who are firmly on the nature side and believe that we as a species all have a language-acquisition device in our brains and there is a Universal Grammar common to us all. Others are sceptical and think nurture has just as important a role.

Whether you're on the nature or nurture side of the argument, or somewhere in the middle, what seems incontrovertible is that no one knows for sure. It's one of the greatest mysteries of our species, but one we're very slowly beginning to uncover.

Talking Animals

Why is it that we can but other animals can't? Let's take our closest relatives, our primate cousins, the gorillas.

Ambam is a silverback lowland gorilla who is being rehabilitated at Aspinall's Port Lympne centre in Kent. Ambam became something of a YouTube sensation when he was filmed walking on two legs. It's extraordinary to see this immense creature standing upright, looking around, just like a human being. Gazing into his eyes, you see intelligence, and emotions. Gorillas can laugh and cry. But there the communication seems to stop. Because what they can't do is talk.

Ambam the gorilla at Aspinall's Port Lympne Centre, Kent

For a start they just don't have the vocal equipment. A defining feature for our species' ability to speak is a lowered larynx. This, combined with controlled breathing (apes cannot hold their breath) and the finely tuned muscles in the tongue and lip, enable the air we breathe to be forced through the vocal chords and into the mouth, where it can be shaped with extraordinary subtlety. Just say 'hello' to yourself and see how your mouth and tongue move. Well, no other primate can do this. And of course we have the brains for it. But why are we so keen to speak to other animals?

Perhaps it's because of our otherness from the rest of creation that we are fascinated by the idea. From countless mythic stories in tribal cultures around the world, from the Serpent in the Garden of Eden to C. S. Lewis' Narnia chronicles, from *The Jungle Book* and Bugs Bunny to Mr Ed, Dr Doolittle and Stuart Little, the talking mouse, book and movie fiction feeds our yearning for creatures who can talk to us. This desire to communicate with the rest of creation has inspired entertainers, scientists and charlatans in equal measure.

Animals with allegedly extraordinary communication skills have been doing good box office for hundreds of years. Signor Capelli's musical cats were billed in 1829 as 'the greatest wonder in England', and Charles Dickens is said to have watched another of his acts, Munito the Wonder Poodle, who could play dominoes, recognize colours and count. Dickens wrote of seeing the act for the first time and being fooled by the dog's 'answering questions, telling the hour of the day, the day of the week or date of the month, and picking out any cards called for from a pack spread on the ground'.

Toby the Sapient Pig

'Toby the Sapient Pig' was introduced to London audiences around 1817 as 'the greatest curiosity of the present day'. According to the billboards, Toby would 'spell and read, play at cards, tell any person what o'clock it is to a minute by their own watch . . . and what is more astonishing he will discover a person's thoughts'.

Needless to say, most of these acts were based on trickery rather than genuine language ability. Dickens watched Munito's performance more carefully a second time and this time he noticed that the dog was choosing cards by smell rather than by sight – the master had daubed them with aniseed.

Toby the Sapient Pig could apparently spell, read and tell time

We're just as intrigued by the possibility of talking animals today. Type in 'Talking animals' on the YouTube website and you'll be overwhelmed by an astonishing array of video clips of talking pets, like Odie the Pug dog, who yowl-whispers a most convincing 'I love you' on command. But away from the trickery and the mimicry and the fabrication, serious scientific attempts have been made to explore to what extent animals have the ability to learn and understand human language.

Dolphins, horses, parrots and chimpanzees have all been the subject of scientific research and debate. Oprah Winfrey's television show in the USA recently broadcast an interview with Kanzi, a bonobo chimp. Kanzi is known as 'the ape who has conversations with humans'. The debate is a heated one. Linguists like Noam Chomsky argue that language is unique to humans, whose brains evolved with special language modifications which no other animal has. According to Chomsky, humans possess a sort of language gene which enables them to give grammatical order to words. Others suggest that, if earlier hominids had facilities for communication, then these adaptations may still be present in the modern ape. Yet another group of researchers argues that some intelligent animals have the ability to learn some of the fundamental characteristics of human language.

The earliest talking animal story to generate serious scientific research was probably Hans the Counting Horse. His owner, a late nineteenth-century German maths teacher called Wilhelm von Osten, believed that animals were much more intelligent than humans gave them credit for. Von Osten decided to prove his theory by teaching mathematics to a cat, a bear and a horse. The cat and the bear were indifferent, but the horse, an Arab stallion called Hans, showed promise. If Von Osten chalked a number on a blackboard, Hans would use his hoof to tap the number out – a chalked number 4 would produce four taps of the hoof. Questions on addition, subtraction, fractions and spelling could all be answered by Hans with the tapping of his hoof. Word of the clever horse spread, and Von Osten began to exhibit Hans in free shows all over Germany. Huge crowds gathered to watch Hans answer questions posed by his master. 'What is the square root of nine?' 'If the fifth day of the month falls on a Monday, what is the date of the following Thursday?' Hans would be asked to spell out words with taps – one tap for A, two taps for B, and so on. His answers were almost 90 per cent correct.

The German board of education assembled a panel of experts to study this equine genius, and in 1904 the Hans Commission, which included two zoologists, a psychologist and a circus manager, reported that it could find no signs of trickery and that Hans' abilities appeared to be genuine.

The investigation was passed on to a psychologist, Oskar Pfungst, who after careful observation came up with a ground-breaking conclusion. Hans the Horse only gave the correct answer when he could see the questioner and the questioner knew what the answer was. Pfungst had watched Von Osten closely and noticed that as the horse's taps approached the correct answer, von Osten's body posture and facial expressions changed. They became tenser, and then relaxed when the horse made the final, correct tap. This relaxing was the cue to Hans to stop his hoof tapping. Thus Hans the horse was shown to be an animal not so much of great intelligence but rather one of great sensitivity to body language. An animal instinct, in fact.

Hans the Counting Horse and his trainer Wilhelm von Osten

Oskar Pfungst's insight came to be known as the Clever Hans Effect – the influence a questioner's cues may exert on their subject, both human and animal.

Project Pigeon during the Second World War was an attempt by American behaviourist B. F. Skinner to develop a pigeon-guided missile. A lens which could reflect an image of the target on a screen was put in the nose of a bomb; a pigeon trained to recognize the target was placed inside the bomb as well and would peck the screen whenever the missile went off target. Project Pigeon never got off the ground, as pigeon pecking was overshadowed by the development of electronic guidance systems.

It was just one of the many projects Skinner developed as part of his belief that all animals can learn and change behaviour and that language is simply an extension of learned behaviour. His best-known invention was the Skinner Box – or the operant conditioning chamber. The box contained one or more levers which an animal could press and one or more places from which food could be dispensed. Skinner would put a rat or a pigeon into the box and showed that the animals quickly learned that they would get food every time they pressed the lever. Skinner asserted that they would only manipulate the lever if they were rewarded for the action, a process he called 'shaping'. He expanded his theory to conclude that human behaviour, including language, is learned from our environment. There is no difference between the learning that takes place in humans and that of other animals. If Skinner was right, and language is learned and not instinctive, then every animal with intelligence of a certain level should be able to learn to talk and can be nurtured to use language.

What Do Dolphins Talk About?

Dolphins have long been recognized as one of the smartest of all the mammals, with their own highly evolved system of communication of clicks and whistles. In the 1960s, American scientists began to study the complex brain of dolphins and analyse how they communicated with each other and, what's more, whether we could communicate with them. It wasn't long before the US military got interested. Imagine an intelligent animal, able to understand and follow instructions, swimming undetected through enemy waters.

In 1964, Dwight 'Wayne' Batteau was funded by the US Navy to develop a man/dolphin communicator. He described the project as 'a program of research intended to determine the feasibility of establishing a language, approaching English, between man and dolphin'. An electronic device called a transphonometer was designed to convert the vowels and consonants of humans into whistles. These sounds were transmitted underwater, and the dolphins reproduced the whistles. Specially trained Navy personnel were able to learn this whistle language and communicate with trained dolphins.

Batteau drowned in 1967 before the project was completed, and since then most of the information about the Navy's work with dolphins has been classified. However, we know that dolphins have been trained to attach explosives and listening devices to enemy ships and submarines. And in the Gulf wars dolphins were used to search the seabed for mines. The man/dolphin communicator may have enabled humans to issue

A US Navy-trained dolphin used to locate mines and torpedoes

commands to dolphins, but there is no evidence that this is more than say a sheepdog's ability to learn and respond to the commands of its master. Batteau's dolphins didn't talk back – in whistles or otherwise.

Researcher John C. Lilly tried to teach dolphins to talk like humans and claimed in the 1960s that he had trained them to replicate the alphabet. Recordings made at his research centre in the Virgin Islands do have a few examples of dolphins apparently echoing human sounds in high-pitched squeaks, but there was no evidence of dolphins actually uttering human words.

Since then, dolphin-language research studies have concentrated on proving that dolphins can understand and interpret human language rather than replicate it, which is a lot more sensible, as the dolphin's anatomy is not suitable for making human sounds.

In the 1980s, a female bottlenose dolphin was the subject of Louis Herman's animal-language studies at the Kewalo Basin Marine Mammal Laboratory in Hawaii. Researchers used a sign language which allowed them to give the dolphins highly complex instructions. For example, 'left basket right ball' asks the dolphin to put the ball on her right into the basket on her left. But 'right basket left ball' means the opposite – put the ball on the left into the basket on the right. The results were published in 1984 in the human psychology journal *Cognition*. 'The dolphins were able to account for both the meaning of words and how word order affects the meaning,' said Herman.

So far the dream of dialogue between man and dolphins remains just that. Dolphins clearly have a complex sound system, but we're a long way off from interpreting the meaning of their clicks and whistles. Never mind conversing with them, we still don't know what they're saying to each other. Is it ontological discussions on the future of the planet or simply 'Let's go get some salmon to eat'? As one language expert comments, 'Their capacity for communication could range from the level of a dog barking all the way to possible talking.'

The most impressive interspecies language experiments have been with primates, our closest relatives. Noam Chomsky and his followers argue that the ability for language developed in humans after the evolutionary split between humans and primates. They point to the ease with which children acquire language. Children, they insist, have an innate propensity for language which primates simply do not possess. According to Chomsky, it was 'about as likely that an ape will prove to have a language ability as there is an island somewhere with flightless birds waiting for humans to teach them to fly'.

In the 1970s Herbert Terrace, a psychologist at Columbia University, brought a baby chimp (playfully named Nim Chimpsky after the famous linguist) to the LaFarge family in New York. Nim was treated just like one of the family. He had seven human siblings and he was carried everywhere for the first year by his surrogate mother, Stephanie LaFarge. He ate human food, wore nappies and clothes, brushed his teeth at night and – somewhat unusually for a child – enjoyed a cigarette and a cup of coffee. Nim was taught American sign language at home and in a classroom built specially at Colombia.

As Nim hit the terrible twos he became too difficult for the

Nim Chimpsky and researcher both signing 'drink'

LaFarges to look after. He was moved to a mansion owned by the university and was cared for by a series of handlers. After four years, Terrace announced that Nim had a vocabulary of more than 100 words, but in 1977, after the chimp severely bit one of his teachers, Project Nim was stopped. When the results of the project were published, Terrace declared it a failure. He said that, while he was watching a video of Nim signing with a teacher, he realized that the chimp was imitating most of the signs but he almost never made a sign spontaneously. Herbert had tried to avoid the Clever Hans Effect but in the end he concluded that Nim and other chimps who had been taught to sign were merely imitating rather than using language. Compared with a human child, Nim rarely added new word combinations and seemed to have no idea of syntax or the elementary rules of grammar. Chomsky, it seemed, had been right. Nim would never use language in the way humans do – using grammar to form sentences and express ideas. He never asked a question.

After the collapse of the project, no one quite knew what to do with Nim, this chimp who had been raised to believe he was a boy.

He'd never been in a cage before or met other chimpanzees. There was a public outcry after he was sold to a laboratory of experimental medicine. Nim's surrogate sister Jenny Lee told a reporter: 'How do you reconcile a tiny chimp in blue blankets, drinking from a bottle and wearing Pampers . . . and then, when he is ten – him in a lab, in a cage, with nothing soft, nothing warm, with no people? This is my brother. This is somebody I raised.'

Nim was eventually offered a home in an animal sanctuary in Texas, where he spent hours looking through old magazines and trying to sign to keepers. Other chimps were introduced as companions, but Nim always relished human companionship. He died of a heart attack in March 2000, aged twenty-six.

The latest celebrity 'talking' primate is Kanzi, a bonobo chimpanzee who has been taught to 'speak' by pointing at lexigram symbols on a computer. On his keyboard are hundreds of colourful symbols representing all the words that he knows. Not just easy words like *ball* and *banana* and *tickle* but difficult concept words like *later* and *from*. When he touches the symbol the word is repeated out loud. What excites primatologist Dr Sue Savage-Rumbaugh at Iowa's Great Ape Trust is that Kanzi is putting two-word sentences together. For instance, the word 'flood' isn't on the lexigram, so when there was a flood in Iowa, Kanzi pointed out two words *big* and *water*. He was given kale to eat and, finding it difficult to chew, he pointed at *slow* and *lettuce*.

Kanzi also seems to have developed theory of mind, a skill closely linked to human language. It means he is able to imagine the world from another person's point of view. When he noticed one of the researchers had a missing finger, he asked 'hurt?' And he uses language creatively. Kanzi was told that a Swedish scientist called Pår Segerdahl was coming to visit and was bringing him some bread. There was no symbol on the lexigram for scientist so Kanzi pointed to the symbols for *bread* and *pear*. When he was asked if he was talking about Pår or pears to eat, he pointed to the scientist.

Researchers have observed that Kanzi makes the same four sounds for four words: *banana, grape, juice* and *yes*. Could this be the beginnings of speech, or is it no more significant than a dog who gives a particular bark when it sees its master? Critics say Kanzi is reacting to body language and that researchers interpret his use of words too creatively. Dr Savage-Rumbaugh has tried to avoid the Clever Hans Effect by conducting some of her tests sitting very still and wearing a welder's mask so Kanzi can't see her face. She makes a series of unlikely requests, like 'put the pine needles in the fridge', which Kanzi almost always understands.

The debate continues. We're certainly not imagining a future world of prattling primates as in *Planet of the Apes*. Perhaps the most useful way to look at it is this: just as Kanzi has turned out to be an accomplished crafter and user of stone tools, including some very sharp knives (which helps scientists understand skills used by our early prehistoric ancestors), so, at the very least, these talking experiments allow us a fascinating glimpse of a stage in the evolution of human language. But in the end the crucial difference is our brains. They are nearly four times the size of a chimp's. It's all about that extra kilo of grey matter.

Grey Matter

With advances in medical technologies, science is at last beginning to penetrate one of the greatest mysteries of nature: that three pounds of mushy grey matter lodged between our ears, the human brain, the bodily organ which you are using to understand these words. Its ability to learn and process language is one of the true wonders of the universe.

Professor Cathy Price works at the Wellcome Trust Centre for Neuroimaging at the University College London medical school, one of the foremost research centres in neuroscience. Her research programme aims to create an anatomical model of how language works. Using structural and functional magnetic resonance imaging (MRIs, those huge doughnut-like machines which scan the body to create a three-dimensional image), she hopes to build a picture of which parts of the brain are used to process language. As is so often the case with the workings of the body, the insights into how they work are most evident when they stop working – jaundice tells us about the function of the liver in a way that a healthy liver simply cannot. Much of her research concerns how speech and reading are lost and recovered following strokes, when different parts of the brain are knocked out.

Cathy candidly confesses that, though she's been studying the brain for a very long time, very little is understood. In fact, she says we are in the process of unlearning and dismantling what we thought we knew as these new techniques for looking into the fissures and folds of the brain are developed.

This classical view of the brain is the one we are familiar with from textbooks. Right, left hemisphere, cerebral cortex, cerebellum, basal ganglia, limbic system . . . It was Paul Broca, a French surgeon, anthropologist and member of the Académie Française, who first noticed that people who had suffered strokes in a certain area of the left inferior frontal lobe had problems with speech production. In 1868, he presented his findings to the British Association in Norwich, and in recognition of his pioneering work not long after it was decided to name this part of the brain Broca's area. A Prussian neuropathologist called Carl Wernicke, reading Broca's research, did his own and found out that his stroke patents who had suffered damage to the posterior left hemisphere had problems with speech comprehension rather than production. So this area of the brain

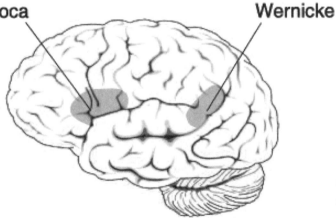

Broca   **Wernicke**

The classical view: the two major language centres of the brain

was called Wernicke's area. For a century and a half these two areas were believed to be the parts of the brain where language production and language comprehension were situated. Research by people like Cathy is now causing that theory to be revised.

What is now evident is that language occupies a far greater part of the overall brain's activity than just the Broca and Wernicke areas – more than half the brain appears to be involved in some speech-related function. There are so many factors: the breathing, the shaping of the tongue and lips, fine motor control of facial muscles, control of the vocal chords, and that's before any of the cognition processes begin. And here's where the mystery deepens. Where does memory reside? How is it tapped? Do thoughts control language, and how do we then process the thought into the language we want? How does that work for bilingual or multilingual speakers?

'I used to think my aim was to be able to understand language in the brain,' says Cathy. 'I've now reached the stage of my life where I don't think it's going to happen within my lifetime, and now I have much more specific aims as to how I can apply the information I have for clinical use. So I've gone from this big ambition – will we ever understand it? – down to much more specific aims that I think might be clinically beneficial.'

Cathy hopes that one day we will be able to have a more meaningful interpretation of how the brain works, but at present we don't have a coherent story into which all the pieces fit. We might be able to colour it, make models of it, say this bit is working when we ask for a cup of tea and this bit when we feel anger or love, but the really big questions concerning the actual physiological nature of the organ are yet to be answered. So far, we've only just scratched the surface.

So have Cathy's researches made her swing one way or another in the whole nature versus nurture debate as to how language is acquired? Her practical investigations into grey matter have made her believe there is a primal language structure in the brain, and it is remarkably consistent in all individuals, across all cultures and irrespective of whether the language is spoken, signed or even read as Braille.

So how, then, did this primal proto-human language emerge?

Noam Chomsky, father of modern linguistics

Noam Chomsky

Noam Chomksy could be said to have invented modern linguistics. Of course, he is much more than this, but a half century of teaching, writing and being on the public stage has done little to reduce the public perception that Noam Chomsky *is* modern linguistics. His ground-breaking theory that *homo sapiens* has an innate language ability, and all grammars and syntax are essentially the same and hard-wired into our brains, remains his major contribution in the field.

Seeing how quickly and easily children learn language without any seeming effort, Chomsky posited that there is a 'Language Acquisition Device' (LAD) which is unique to humans. It's this LAD, deep within our brains, that enables us to speak, whereas kittens or dormice are mute. According to Chomsky, there is a deep-set 'Universal Grammar' common to every human language.

Chomsky's work with linguistics waned as he became more outspoken in his criticism of American foreign policy during the Vietnam War in the 1960s. His political activism (he calls himself a libertarian socialist or anarcho-syndicalist) has made him one of the most controversial intellectual figures in the USA. His linguistic theories, too, have divided opinion.

How We Learned to Talk

Dr Michael Tomasello is the Florida-born co-director of the Max Planck Institute for Evolutionary Anthropology in Leipzig, Germany. He divides his time between their spanking new building near the university and Leipzig zoo's splendid open-plan primate enclosure.

Tomasello's studies with the great apes and how they communicate and interact socially with each other have expanded in recent years to the observation of children and how they acquire language. His findings amongst the apes and children have led him to reject the idea of a language gene or a human instinct for language. He is at odds with the likes of Noam Chomsky, arguing that human language evolved out of a need for social interaction. 'Human communication is a fundamentally cooperative enterprise,' he says, and the first signs of this communicating which you see in young children are the two basic types of human gesture: pointing and pantomiming. Pointing is arguably the most fundamental type of human gesture. Tomasello describes this scenario: You're standing in a line. The line has moved forward, and a man hasn't noticed because he's talking to the person behind him. Someone from still further back points him to the newly opened space. There's a universal understanding of what is meant by the pointing: 'Hey, there's a space. Move up!' Tomasello illustrates the other type of basic human gesture, pantomiming, or signing for the imagination, as he calls it, with another scenario: You're at the front of the lecture hall, ready to give a lecture. A friend in the audience fiddles with her shirt button, frowning at you. You look down at your own, and, sure enough, it's unbuttoned. Tomasello argues that these simple gestures are actually a form of complex communication; our pointing and miming to each other is a kind of cooperation that is unique to humans.

Experts call this *shared intentionality* or *recursive mind reading*. In simpler terms, it's a sort of unspoken shared goal to collaborate, an awareness of what the other person is thinking. For instance: We're sitting beside each other; I'm looking at a book on the table; you're looking at the book; I know that you're looking at it and I know that you know that I know. Apes apparently don't think like this. The theory is that conventional languages – first signed and then vocal – developed from these natural gestures of early man. Language evolved because humans needed it to survive and prosper; they needed to be able to do more than gesture – to organize themselves, to pass on knowledge, to gossip, to plan. Tomasello doesn't reject the idea that the ability to speak, to produce sounds that apes can't articulate, may be genetic, but he argues that the reason we evolved speech was to allow us to communicate *more effectively* – so the communication, the awareness of shared cooperation, was already there.

Man's first language, according to Tomasello, would have been a signed one. It's the natural progression from pointing and pantomiming. He explains it with another scenario. Imagine two groups of children who have never communicated with anyone. Each group is isolated on its own desert island, one group with their mouths covered with duct tape, so unable to talk, the other with their hands tied behind their backs, so unable to gesture. We know what would happen with the first group, because it's what happens to deaf children who are brought up in families where no one knows sign language. When these children come together they develop a sophisticated sign language with grammar and syntax. We don't know what would happen with the children whose hands have been tied, but Tomasello thinks it unlikely that they would have been able to invent their own structured vocalization. They might have emotional responses – screams, howls, etc. – and be able to mimic sounds, like apes, but their sounds would not be language.

Looking at how children develop allows you to witness how, through the awareness of shared cooperation, they quickly move from gesturing to talking. In evolutionary terms could they be at the stage that early man was? And if so, what propelled the pointing early man to make that leap forward to fully fledged language? Well, it would have been an immense evolutionary advantage for a group to develop the speech skills necessary for the creating and passing on of knowledge. Whether it was language itself or other cognitive functions that propelled *homo sapiens* forward is one of those chicken-and-egg questions that keep recurring

FoxP2 – the Language Gene

The Holy Grail for those who argue that humans are born with an innate language ability must surely be scientific proof that we are genetically wired to communicate in a way that no other animal is. Scientists now believe they've found the answer – well, part of the answer – with the discovery of a so-called language gene. In London in the 1990s, a family known simply as the 'KE family' came to the attention of researchers at Oxford University. Over half of the thirty-seven family members, extending over three generations, were born with severe speech and language difficulties. They struggled to speak grammatically correct sentences and they couldn't move their tongues or jaws properly when they tried to speak. When the researchers examined the unaffected members of the family as well as the affected ones, they identified a single gene as the apparent cause of the speech problems. The FoxP2 gene, which stands for Forkhead Box Protein P2, is a *transcriptional regulator* – in layman's terms, it switches genes on and off during the creation of a body's tissue. All the members of the KE family who had speech problems had a damaged FoxP2 gene. As further evidence, the researchers identified a boy, known as CS, who was unrelated to the KE family but had similar speech and language problems. He too had the faulty FoxP2 gene.

Human FoxP2 appears to affect tissue growth in the brain, giving us the ability to shape very precisely the sounds we make with our mouths. Two copies of the gene are needed, one from the mother, one from the father. A fault seems to have developed at the embryo stage of those members of the KE family with speech problems. They had only one copy of the gene, and this fault distorted the normal genetic sequencing required during the early development of the

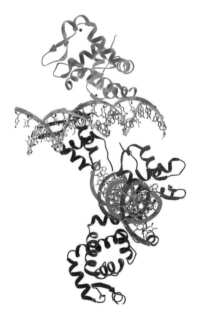

FoxP2 gene (Forkhead Box Protein P2) 'the Language Gene'

regions in their brains associated with speech and language.

So does this mean humans are the only species with the FoxP2 gene? Have Chomsky et al. been proven correct? Well, no. The gene has been identified in all mammals. In fact, the only difference between human FoxP2 and chimp FoxP2 is two amino acids; with mice, it's just three. FoxP2 affects birdsong: zebra finches are just seven amino acids away from humans. If you take away the gene, the birds can't learn their songs. It's even thought to help bats with echo location. What researchers now speculate is that those two amino acid differences between humans and chimps led to the evolution of language in humans whilst chimps remained language-less. FoxP2 was not necessarily the gene which gave humans language, but it certainly influenced humans' ability to speak, enabling anatomical changes such as the lowering of the larynx and development of fine motor movements of the mouth and the lips.

In the labs at the Max Planck Institute in Leipzig, scientists have been working with mice to calculate at what point in evolution the specifically human version of the FoxP2 gene developed. Dr Wolfgang Enard believes that human FoxP2 developed 120,000 to 200,000 years ago, at roughly the same time as the emergence of *homo sapiens*. A variation in the gene of just two amino acids could have spread through the generations to the entire human population. The genetic sequencing of Neanderthal Man has now been completed and, rather thrillingly, he too had the same FoxP2 gene as *homo sapiens*. This doesn't mean that 30,000 years ago (before they died out) Neanderthals were enjoying their version of a Homeric epic round the fire, but, according to Dr Enard, it does suggest that they may have communicated with a more elaborate system of grunting than first thought. The discovery of a hyoid bone in a recently uncovered Neanderthal skeleton, crucial in the vocal apparatus of *homo sapiens*, means there's no reason why they should not have been competent linguists. But we'll never know for sure. What is certain is that the mapping of the human genome is in its infancy. Dr Enard reckons there could be anywhere between 10 and 1,000 language genes. 'This is hopefully the first of many language genes to be discovered,' says Enard. 'It is compatible with the hypothesis that language could have been the decisive event that made human culture possible.'

'All Gone Sticky'

Steven Pinker is one of the foremost linguistic experts in the world, author of many bestselling books on language and a professor at Harvard University's Department of Psychology. He is, without doubt, the most high-profile, and some would say most controversial, disciple of Noam Chomsky.

Pinker has built on some of Chomsky's theories to create his own view on linguistics. His work is more technical and less abstract than Chomsky's; rather than looking at the deep structures within the mind, he is fascinated by the surface manifestations – how a child learns to speak, why complex grammars are useful and what part natural selection plays.

'Language at a bare minimum needs words,' explains Pinker, 'and the words have to be the same words that everyone else is using. If you had your own private language, if it were even possible for language to just spring out of the brain, it would be completely useless, no one would understand a word you're saying. So the child has to be attuned to the words that are floating around in the linguistic environment and to the sound patterns and to the rules of grammar that order them in meaningful sequences.'

We usually think of that input as coming from parents, although children are more attuned to their peers than to their parents, as any immigrant will attest. If you move from Britain to America and your children grow up with American children, they will end up with an American accent. So there is an environmental component, but, says Pinker, 'there also has to be some kind of talent in the child's brain that allows them not just to parrot back the exact words and sentences they've heard. It'd be very upsetting if that's what your child did. We expect children right from the beginning to compose their own sentences, to abstract the rules of combination, the rules of grammar, so that they can talk about new events and new thoughts and take the familiar words but rearrange them in new sequences.'

So can a young child actually come up with a sentence that's entirely original, perhaps never before uttered in the history of their language?

Steven Pinker, one of the world's leading linguists

Pinker says yes, right from the beginning, from the time when children first start putting two words together, some of those combinations are clearly from their own creativity. An example: a child whose hands are covered with jam and wants his mother to wash them. The mother washes off the jam, and the child says, 'All gone sticky.' Pinker explains: 'Now that doesn't correspond to any adult English sentence, but the child had those three words and had the formula that put them in that order to express the idea of the passing of a state. So that's already involving, almost without it being a tense, a sense of past: "gone". And so, even without apparently using a sophisticated series of words, there was a sense of saying something once happened, and it's now over. There's the concept of a change of state, and moreover there's the ability to put words together in brand new sequences that had not been previously memorized but that express a coherent thought.'

He gives another example: 'more outside', i.e. a child wants to be taken out.

'That's quite a cognitive feat to have the concept of addition or repetition which had been associated with the word *more*, and to put those two words together in a way that doesn't exactly map on to adult syntax but that an adult can certainly comprehend. And that combination, that little baby micro-sentence, came from the linguistic creativity of the child's mind. The mother or father would never have said "more outside", but the mother would often have said "more", and the baby associates "more" with more food, but to transfer that to something as abstract as the outside is an act of language no matter how primitive it seems.'

Pinker elaborates by saying that, right at the beginning of language learning, when the first pairs of words are being put together, we use words in conjugational and inflectional forms, like when a child says, 'He sticked it on the paper,' or 'He teared the paper,' or 'We holded the baby rabbits.' We know these forms haven't simply been parroted from adults, because adults don't put regular past-tense endings on irregular verbs like *stick* and *tear*, but the child has clearly abstracted a cognitive equivalent of a rule of grammar: to form the past tense, add 'ed'. So that allows the child to come out with forms that are new for the child and haven't simply been memorized from what they hear from their parents or peers.

The essence of language is to combine things according to rules. So you stick a subject and a predicate together and you get a sentence. This combination can be done in sequence. So you can talk not just about the *rat* but also the *rat who ate the cheese*. And you can talk not just about the *cat* but the *cat who killed the rat who ate the cheese*. But you can also stick one phrase inside another. Take this example: *The boy that I saw the day that I was having my teeth done left*. And then when you put a noun phrase inside a noun phrase you get *the boy's mother* or, adding still more, *the boy's mother's friend*, which is a noun inside a noun inside a noun. So you get: *The boy's mother's friend that I saw the day that I was having my teeth done left*. 'It's an extremely powerful information-processing technique,' says Pinker 'because it means with a very small tool kit you can do an infinite number of things. In the case of language, what it gives us is the ability to talk about an unlimited number of ideas using the fixed vocabulary that we acquired at our mother's knee.

The Wug Test

One of the early experiments that aimed to see just how children acquired these grammatical rules without seemingly trying was conducted in 1958 by one of Chomsky's contemporaries, Jean Berko Gleason, now Professor Emerita at Boston University.

A sprightly woman in her seventies, Jean still has a passionate curiosity about how language works, though with fewer answers and less certainty than when she started out more than fifty years ago. She has brought her original Wug cards with her – simple, hand-drawn pictures of an imaginary animal. Jean tells a group of children aged between two and four that this animal is called a *wug*; then she shows a picture of two wugs to see if they can make the connection and make *wug* into the plural *wugs*. Some do, some don't. She tries other animals and then action pictures like a man balancing a kettle on his head to see how the children cope with verb tenses of an imaginary verb. The interchange with one of the children, Julian, is revealing:

Jean: Okay, okay, now this is another creature, this one's called a tass. That's a tass.

Julian: Yeah, okay.

Jean: Now, there's another one, there are two of them. So there are two.

Julian: Tass.

Jean: Tass.

Julian: Tass.

Jean: Two tass. Okay. Okay, very, very good. Okay. This is a man who knows how to spo. He is spoing. He did the same thing yesterday. Yesterday, what did he do, yesterday he... ?

Julian: Um, balanced the, a kettle on his head.

Jean: Well, yes he did, but he was spoing yesterday so

THIS IS A WUG.

NOW THERE IS ANOTHER ONE.

THERE ARE TWO OF THEM.

THERE ARE TWO _____.

The Wug Test

yesterday he . . . ?

Julian: Spoed.

Jean: Perfect. Very, very good. He spoed. Okay. Now, this is a man who knows how to gling. He is glinging. He did the same thing yesterday, what did he do yesterday, yesterday he . . . ?

Julian: Glinged

Jean: Glinged, very, very good. Very good.

Jean carries on for half an hour, and it's clear there's quite some variation in the children's ability to form these grammatical structures. She explains that much depends on the individual child's development; a few months makes a huge difference in a two-, three- or four-year-old.

So what is Jean's take on the whole nature/nurture debate? She believes that what makes us human beings is our capacity to build our brains. 'Young kids' brains are not formed when they're born, there's not some organ in them that is grasping language. They have areas that will ultimately be specialized for language but it's through experience, it's through hearing language, it's through interacting with people who use language with you that you build the language in your brain, because you have that capacity.'

Feral Children

If you were to take a healthy newborn baby, nothing wrong with it, physiologically able to hear and speak, and then you were to shut the baby away from all human contact, what language would the baby speak when it uttered its first word? Would it be that putative language that Adam and Eve spoke before the Tower of Babel and the unleashing of thousands of languages? Or gibberish? Or would it speak at all?

At various times in history, experiments have been carried out to isolate children from society, with the express desire of finding this original language. The idea horrifies us now, but in times of despotism and a somewhat laissez-faire approach to human rights, such experiments were possible. The Greek historian Herodotus wrote about Psammetichus I, ruler of Egypt in the seventh century BC. The pharaoh wanted to settle the question of who were the oldest people of mankind, so he gave two newborn babies to a shepherd and ordered him never to speak to them. The hope was that the first word either of the children uttered would be in the root language of all people. One day the children held their hands out to the shepherd and cried 'bekos'. The pharaoh was informed that *bekos* was what the people of Phrygia – modern-day Turkey – called bread and so he decided that the Phrygians were the oldest people in the world.

A monk in the thirteenth century recorded a language experiment of the Holy Roman Emperor Frederick II in which all the infants appear to have died. According to the *Cronica* of Franciscan friar Salimbene di Adam, the emperor bade

> *foster-mothers and nurses to suckle and bathe and wash the children, but in no ways to prattle or speak with them; for he would have learnt whether they would speak the Hebrew language . . . or Greek or Latin or Arabic, or perchance the tongue of their parents of whom they had been born. But he laboured in vain, for the children could not live without clappings of the hands, and gestures and gladness of countenance, and blandishments.*

Scotland's scientifically curious monarch James IV conducted a similar experiment in 1493. According to the historian Robert Lyndsay of Pitscottie, King James sent two infants to be raised by a deaf and dumb woman in a cabin on Inchkeith Island in the middle of the Firth of Forth. James too was searching for the original language of man. There's not much information on the results of the experiment. 'Some say they spoke good Hebrew,' reported Lyndsay; 'for my part I know not, but from report.' The novelist Sir Walter Scott took a more sceptical view when he recounted the tale 300 years later. 'It is more likely they would scream like their dumb nurse, or bleat like the goats and sheep on the island.'

It's unclear what the Mughal emperor Akbar the Great was up to with his language-deprivation experiments in the sixteenth century. Some say he wanted to find out whether people were innately Hindu, Muslim or Christian. Others that he was testing his hunch that babies raised without hearing speech would be unable to speak. He ordered twelve infants to be raised by mute nurses in a house where no speech was ever heard. Several years later, when he visited the children, he found his second hypothesis to be correct. None of them could speak; they communicated instead in signs.

There are probably a few tall tales mixed in with the accounts of these horribly cruel experiments, but generally the results were incon-

Akbar the Great experimented on children with language deprivation

clusive and irrelevant. Clearly there was no 'original' language which infants deprived of speech would speak; they simply didn't speak at all. Stories about feral children, where language has been deprived by chance rather than by the order of a king, are better documented. A feral child is defined as a human child who has lived away from human contact from a very young age and has had little or no experience of human care, loving or social behaviour and, crucially, of human language. Studies of such children can help to understand the process of how language is acquired – how much of it is learned and how much is genetic. Unfortunately, in almost all the studies of feral children, practically nothing is known about the child's life before their capture.

One of the most famous and well-documented cases is the so-called Wild Boy of the Aveyron. In 1797, in the midst of the madness of the French Revolution, a boy (aged about eleven or twelve) was discovered wandering in the forests of France's south-eastern Massif Central, one of the most rugged and least inhabited regions of Europe. A hunter came upon the lad living alone, without clothes, without tools, eating and defecating like an animal and, most significantly, without language. He was captured and brought to Paris, where crowds thronged to see him. Victor, as the wild boy became known, offered France an opportunity to see the romantic theories of Jean-Jacques Rousseau in practice. Rousseau's famous treatise posited the development of a child away from the influences of society, so that he could grow up as Nature intended. What the sightseers found in reality was a dirty, grunting creature, rocking backwards and forwards like an animal in a zoo.

Victor ended up under the care of a young doctor, Dr Jean-Marc Itard, at the recently established Institute for Deaf Mutes in Paris, which had been founded by the celebrated inventor of sign language Abbé Charles-Michel de l'Epée. Itard was a child of the Enlightenment and was fascinated by Locke's theory that we are born with empty heads and that our ideas arise from what we perceive and experience. Itard decided he would need to educate Victor from the very beginning and so became his Pygmalion. Over the next five years he devoted himself to the boy, teaching him how to eat, use a toilet, repress his animal urges (particularly with the female inmates)

Victor, the Wild Boy
of Aveyron

and learn French. Victor's vocal chords, like any muscle unused for so long, needed to be exercised, and Victor gradually began to articulate sounds, rather in the manner of a baby. It became clear he was not a deaf mute, although he may have suffered to some degree from autism which may have been the reason he had been abandoned in the first place. Itard's methodology is what we would now consider behaviour modification – a system of punishment and rewards; with this treatment he hoped to prove one of the tenets of the French Revolution, namely that nurture could modify nature. At the end of five years Victor had learned some basic signs but, critically, he never learned to speak. Itard gave up, and Victor, after his brief moment of celebrity, lived in anonymity with Itard's housekeeper until his death in Paris in 1828.

By a strange coincidence, a century and a half later, the premiere of François Truffaut's film about Victor, *L'Enfant sauvage*, was showing in Los Angeles at exactly the same time as the discovery of America's most notorious feral child. In December 1970, authorities in Los Angeles found a thirteen-year-old girl – later given the pseudonym Genie – tethered to a potty chair. She had spent her life since a baby naked and locked in her bedroom. Her father had forbidden his wife and son to speak to her and had barked and growled at her like a dog to keep her quiet. She had been fed a diet of milk and baby food and wore a nappy. When Genie was found, she couldn't straighten her arms or legs and didn't know how to chew or control her bladder or bowels. And she was almost entirely mute.

Genie became the subject of intense interest amongst linguists and behaviourists. Here was a young teenager who had missed out on the crucial stages of language development as a child. Studying Genie might finally answer questions about whether we have an innate ability to speak or whether we learn from our environment, and whether we can learn language at any time or only when we are young. Over the next six years Genie had intensive language training and testing. She learned some words and was good at non-verbal communication but she was never able to put words into a logical order ('applesauce buy store') and construct meaningful sentences. Despite evidence of an innate intelligence, she spoke at about the level of a two-year-old: 'want milk'. Her lack of progress seemed to prove Chomsky's theory that the innate left-brain capacity to learn language must be developed before the onset of puberty or it becomes unable to function for language acquisition.

Genie was fought over by psychologists, linguists, social services and her natural mother. But in the end, as she remained unable to master the basics of language, the researchers lost interest in her. It was the abandonment of Victor all over again. Today Genie is believed to be living in an adult care home somewhere in Southern California. She is still unable to speak in meaningful sentences.

Sign Languages

It was no coincidence that Victor, the wild boy of the French Revolution, landed up in the newly established hospital for deaf mutes in Paris, where for the first time sign language was being taught (using the system created by Abbé de l'Epée half a century earlier). What is apparent with Victor – and with Genie – is that the ability to talk is not the same as the ability to communicate using language. Speech is only one part of the complex thing we do when we communicate. Sign language – using visual signs rather than patterns of sound to express our thoughts – provides a different insight into how language evolves.

There are hundreds of sign languages around the world. From Bolivia to Finland, Somalia to Brunei, intricate systems of hand shapes, body movements and facial expressions exist wherever there are deaf communities. There's even international sign language – Gestuno.

Sign languages develop independently from their oral counterparts. The British and Americans may speak a common English but they have two quite different sign languages. Isolated deaf communities tend to evolve their own unique languages. Early settlers to Martha's Vineyard off the coast of Massachusetts carried a gene for deafness. Soon there were so many deaf people on the island that the inhabitants developed their own sign language; by the nineteenth century hearing people moving into the area had to learn sign language in order to live in the community. The language eventually merged with mainland signs to form the American Sign Language. Yucatec Maya is signed in isolated villages in south central Yucatan in Mexico; Kata Kolok is the language signed in two neighbouring villages in northern Bali, Indonesia. And 150 members of the Al Sayyid Bedouin tribe in the Negev Desert of southern Israel (where the rate of deafness is fifty times the norm) have a sign language with a unique grammatical and linguistic structure uninfluenced by local Arab or Hebrew spoken language patterns.

The first historical record in Britain of signing was of a Leicester wedding ceremony on 5 February 1576 between one Thomas Tilsye,

BRITISH FINGERSPELLING ALPHABET RIGHT-HANDED VERSION

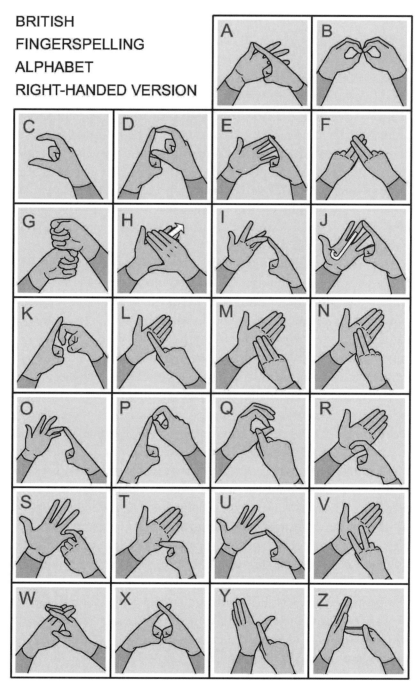

Sign language varies from country to country

deaf and dumb since birth, and Ursula Russell. Thomas made his wedding vows to his intended by a series of gestures, including the tolling of a bell for 'till death do us part'. As the vicar of St Martin's Church wrote in the parish register:

> *First he embraced her with his armes, and tooke her by the hande, put a ringe upon her finger, and layde his hande upon his harte, and upon her harte, and helde up his hands towards heaven; and then to shew his continuence to dwell with her to his lyves ende, he did it by closinge of his eyes with his hands, and diggine out the earthe with his fote, and pullinge as though he would ring a bell.*

Thomas Tilsye's hand gestures were not sign language; his mimicking of actions and pointing to objects do not constitute a true language with its own syntax and grammar. But other anecdotal records from the sixteenth century onwards reveal evidence of simple deaf languages evolving in deaf communities in Britain. In his 1602 *Survey of Cornwall*, Richard Carew wrote about a young deaf man, Edward Bone, the servant of the local MP. Bone could lip-read but not speak. Carew describes how he could relay information to his master with an elaborate system of gestures but later, when he met up with his a deaf friend, he used a completely different signing style.

> *[The] two, when they chanced to meet, would use such kind embracements, such strange, often, and earnest tokenings and such hearty laughters and other passionate gestures, that their want of tongue seemed rather a hindrance to others conceiving them than to their conceiving each other.*

The diarist Samuel Pepys described having dinner with his friend Sir George Downing (after whom Downing Street was named) on 9 November 1666, when a deaf servant had a conversation in sign language with his master about a fire in Whitehall: 'there comes in that Dumb boy that I knew in Oliver's time, who is mightily acquainted here and with Downing, and he made strange signs of the fire and how the King was abroad, and many things they under-stood but I could not'. Downing had been to school in Kent, in a community where congenital deafness was rife. Could he have learned Old Kent Sign Language, an extinct deaf sign language thought to have existed in the county at the time? Was his servant Kentish or did Downing perhaps teach OKSL to him?

The written history of sign language began in the seventeenth century in Spain. In 1620, a priest, Juan Pablo Bonet, published a ground-breaking book, *Reduction of Letters and Art for Teaching Mute People to Speak*. In it he set out a system for using a signed alphabet to teach deaf people. A century later, the French priest Abbé Charles-Michel de l'Epée set up the first public school for deaf children. Abbé de l'Epée had first come across deaf people when he was invited by a parishioner to administer Communion to her two daughters. He apparently mistook the silence of the two little girls for rudeness; when he found out they were deaf, he was inspired to find out more about how other deaf people communicated. He observed the rudimentary signing being used by the sisters and by

Abbé Charles-Michel de l'Epée set up the first school for the deaf

other deaf people in Paris; from that – and drawing upon Bonet's signed alphabet – he developed a standardized system of sign language which he then taught in the world's first school for deaf children.

L'Epée's method, '*signes méthodiques*', combined a system of gestures with other invented signs which represented all the verb ending, articles, prepositions and auxiliary verbs of the French language. It was difficult to learn, and deaf students reportedly preferred to use their own community's signing outside the school. However, the method quickly spread throughout France, Europe and beyond. One of L'Epée's disciples, a man called Lauren Clerc, moved to Hartford, Connecticut, in 1817 to help establish sign language in the New World and created American Sign Language (ASL for short).

Established signed languages, just like spoken ones, can still struggle for their survival. In Britain, for example, deaf children were forbidden to use sign language from the 1880s onwards and had to learn to lip-read instead. It wasn't until 1974 that it was agreed that British Sign Language is a language in its own right, although even today it has no legal status.

A theatre for and by deaf people might sound a bit strange but of course mime has been with us since the mimetic arts first started. Even if you don't understand the detailed language of the signing, the expressiveness of so many of them gives you a good clue as to what is happening. Nonetheless, the fact that the National Theatre of the Deaf (NTD), based in Connecticut, has been in existence since 1967 is a testament to the quality of their work. In the 1970s, NTD was the darling of the experimental and avant-garde, with directors like Peter Brook coming to experience and learn how the emotional directness of signing could be harnessed for dramatic effect.

Janiece, Ian and Claudia are part of the travelling NTD group and spend a good part of the year performing in schools and teaching children the basics of ASL. Janiece is the talking member of the team and translates for Ian and Claudia, who only sign. Claudia is from Germany so has a different, German sign language (which doesn't put the verb at the end as in spoken German!).

Their sign language is a combination of spelling, gesture and cultural shorthand, which is perhaps what makes it so appealing for a theatrical performance. As an example they make the sign for

National Theatre of the Deaf, Connecticut

Madonna, which is unmistakably that of pointy breasts, and then for Bill Clinton, which incorporates the sign for a sexual cheat. Like all languages, it seeks both clarity and efficiency, but on top of that can include character and reputation in a witty manner that dull spelling cannot.

Janiece agrees with that: 'The deaf tend to put more of the spirit of something into the language – a form of bio-linguistics.' The show itself is a wonderful experience. The actions are so dynamic and tell so much of the story that Janiece barely needs to translate the dialogue. It's energetic and funny, and it's easy to see why Peter Brook, theatrical magpie that he is, would want to include some of their techniques in his productions.

One of the most curious cases of sign language was uncovered in Nicaragua in Central America. In June 1986, American linguist Judy Kegl received an unusual phone call from the revolutionary government of Nicaragua. The education minister explained his country had a problem. They had opened two new schools for deaf children and were trying to teach them sign language. The trouble was, the children refused to be taught and were instead miming to each other with their own home-made gestures. Their teachers couldn't understand what they were saying, and the children couldn't understand the teachers. Would Judy fly down from Boston and advise the teachers what to do next?

Judy, a sign language expert from MIT, arrived in Nicaragua expecting to find the children communicating to each other with a jumble of basic gestures. Instead what she found astonished her. The children, most of whom had never met each other before and who had never been taught any form of signing, had compared the signs they had been using at home, shared them and modified them. They had added verb agreement and other grammatical conventions. They had, in fact, developed their own complex sign language with precise rules and order – now known as Idioma de Senãs de Nicaragua (ISN). It's akin to the process in spoken language when pidgin – spoken by a community who have a variety of different languages and devise basic vocabulary to be able to communicate – evolves into creole when it's passed on to the next generation for whom it is their first language. The Nicaraguan children seemed to be developing the equivalent of a signing creole, and, as with creole, the crucial transformation from crude gesture to language was syntax.

All human languages are governed by complex rules, by syntax. As Judy says, 'Syntax, the constraints on language, is something all human beings share. Constraints that are imparted to us by the fact that we share a single human brain . . . the ability to organize information, that allows us to construct novel sentences that have never been said before; that allows us to tell a story, to prophesy, to lie. I can surely communicate for communication's sake when I have syntax; then I can truly use a language.'

One child who learned the invented language at school was Adrian Perez, who now teaches Idioma de Senãs de Nicaragua himself. He describes learning ISN as 'like a rocket going off in your head'. Without language, 'you can't express your feelings. Your thoughts may be there, but you can't get them out. And you can't get new thoughts in.'

Judy adopted two of the Nicaraguan deaf children, and they now live in the USA. They sign with great animation, using the face as much as their hands.

Judy talks about the four instincts that bring us to language. The first is rhythm. 'The single gesture doesn't have rhythm. But you bring a whole bunch of gesturers together, and they repeat, and you create an environment in which communication has the kind of rhythmicity that draws instinct.' The second, she says, is 'monkey see, monkey do', which is about aping someone. In fact she admits that the term is totally wrong, because monkeys don't ape, but primates do, so it should really be 'ape see, ape do', but that doesn't sound nearly as good. Number three is the language instinct: we are born with these expectations for what a language is and what a language should be like. And number four she calls peer pressure – the fact that the very young child thinks that what they believe to be the grammar of the language is what everyone believes. They get to be corrected. They become like the people around them.

So does sign language, which, after all, is as proper a language as any other, have the same critical period for learning as spoken language? Normally up until puberty any language can be learned with ease, but after that it's a much harder task.

'If you're beyond the critical period,' says Judy, 'I can expose you to language until the cows come home, and you're not going to acquire it the way that a native speaker would.' If you're before the critical period for language, she adds, the beauty is, all you need is the compulsion to copy and to be like others.

Such are the expectations of what language is that what is in effect a series of gestures can become a language. So the young children in that first generation in Nicaragua, who came into an environment with older children who were just gesturing, using a whole mish-mash of communication, not language, were fooled into thinking that the gestures were a language; then their expectations kicked in and actually turned it into a full-blown language. The heart of it is language always comes from our expectations about what language should be. Most of the time our expectations come to fit in with the expectations of the adults around us. In the Nicaraguan case the language instinct created a whole new sign language.

It's tempting to liken the pre-signing children to pre-language *homo sapiens* thousands of years ago. Was the moment when a group of Nicaraguan children began to form their own sign language similar to a time long ago when a family group of early humans – perhaps sitting round a fire at night – began to share stories with each other using signs which followed an agreed pattern? Judy Kegl is excited by the idea.

'You know, we can look back and we can theorize how language came into being, but we can't really go back and see it happening. This is a case where we were actually able to sit there during the time that a language was coming into being, and look at it, and see the process while that was happening.'

And so to the big question: has Judy's experience in Nicaragua tilted her one way or another in the schism, or maybe one should just say controversy, that has bedeviled linguists since Chomsky: the idea of a linguistic hard-wiring rather than language being something we acquire?

She is unequivocal: 'If you have children you've had the experience. You watch your children learn English. If you really watch them, they're learning at a much more rapid pace and they're doing a lot more with it than you ever could teach them individually. So in Nicaragua there was no English around. There was no Spanish around. There was no sign language. It's going to be very hard to convince me that something as complex and rich as any other human language didn't come out of the human brain. But I'm also convinced that language needs a trigger.'

The trigger, whether thousands of years ago sitting round a fire or coming together for the first time at a school in Nicaragua, she believes, is community.

'Eskimos Have a Hundred Words for Snow'

Every culture, whether it be a community of deaf mutes or the Eskimos (using that word as a collective noun for all the indigenous peoples of the Arctic region), develop their own way of describing the world around them. For the most part this is done through language. So does our language alter the way we think?

Urban myths are fascinating things, a mixture of Chinese whispers and a desire to astound and astonish one's friends. One old chestnut is the belief that Eskimos have hundreds of words for snow. It is, of course, untrue, but strip away the exaggeration and you'll find the kernel of a linguistic debate which has exercised minds for centuries. Does the language we speak control how we think? Does the (erroneous) fact that a group of people have multiple words for snow mean that they view snow differently from, say, English-speakers?

Charlemagne, the Holy Roman Emperor, declared that 'to have a second language is to have a second soul'. We talk about a nation's language reflecting their temperament and therefore being more suited for certain subjects: French, flowery and romantic for love; English, practical for trade; and German, logical for science. (Wittgenstein said he was once asked whether Germans think in the order they speak in or think normally first and then mix it up afterwards.) Frederick the Great of Prussia had a more specific set of hypotheses: 'I speak French to my ambassadors, English to my accountants, Italian to my mistress, Latin to my God and German to my horse.'

There is, inevitably, a linguistic theory attached to the debate and, like so many theories surrounding language, this one is unprovable. It's also fiendishly complicated, full of those impenetrable words that academics bandy around, and it's got a name: the Sapir–Whorf Hypothesis. Sounds like something out of *Star Trek*, doesn't it? The hypothesis was developed by two anthropologists – Edmund Sapir and Benjamin Whorf – who conducted extensive research among the native people of North America in the early and mid twentieth century. Whorf concluded that language 'is not merely a reproducing instrument for voicing ideas, but is itself the shaper of ideas . . . We dissect nature along lines laid down by our native languages.'

It was an article by Whorf published in 1940 that prompted the Eskimo words-for-snow myth. What Whorf actually said was that the Eskimos, or to be precise the Inuit people (Yupik and Aleut are also spoken by the various Eskimo tribes) have seven distinct words for snow and must therefore think differently about snow. The initial data was flawed and then extravagantly misinterpreted. In the popular press the seven became fifty, and by 1984 an editorial in the *New York Times* expanded the number to 100. The misinterpretation lies in the grammatical structure of the Inuit, Yupik and Aleut languages. They are 'agglutinated' languages (from the Latin meaning 'to glue together'), where suffixes can be added on to a root word to make a new meaning. It's a kind of synthetic process whereby new words are formed by adding morphemes (linguistic units) together. The number of words are almost limitless, as you can just keep piling on the adjectives to form a new word from the root. So for instance the root word for snow is *kamiktshaq*, and you can add *tluk*, meaning 'bad', to create a new word, *kamiktshaqtluk*. In theory you could add 'pee' and 'walrus' to form a new word which would be the equivalent of 'badsnowpeedonbywalrus'. We do the same sort of thing in English when we add *-(e)s* to words to make plurals, and in compound words such as *shamelessness* or, in an extreme case, *antidistestablishmentarianism*, but normally we make phrases rather than very long words. In fact, English and Eskimo have about the same number of root words for snow – think: *snow, blizzard, sleet, slush, hail, flurry*.

Here's another example. The Pinupti Australian Aboriginals have a word *katarta*, which is the hole left by a goanna when it has broken the surface of its burrow after hibernation. That's a lot of words in English to translate one Pinupti word. It's not in our vocabulary because we don't have goannas and we don't live as closely with nature as the Pinupti. If we did, we'd undoubtedly have invented a word. Likewise, if, like the Inuit, our way of life, our very survival, were determined by snow, we might very well come up with different terms (like a word for the 'wrong sort of snow' which falls on roads and runways and railtracks and brings Britain to a standstill for a week).

Messers Sapir and Whorf also noted that the North American Hopi tribe have two separate words for water – one for water in a container and another for water in an open space – like a pond or a river. Does that infer they have a different way of perceiving water? Hardly. But in English we use the same word for water in a river, water in the sea, water in a glass. A Hopi might be justifiably curious why we English-speakers don't have more distinct words for the stuff.

Perception of colour throws up some interesting conundrums, too. For instance, Russian has more individual words for the different shades of blue than English, so does that mean Russians think differently when looking at a Chagall painting than British people do? The Japanese used to use the word *ao*, which spans both green and blue. When the first traffic lights were introduced into the country in the 1930s the green-coloured go light was called *ao shingoo*. Over the years popular use of *ao* changed to represent mostly blue shades, and *midori* became the popular word for green . . . which made things a bit difficult for the traffic light. So what did the Japanese do? Instead of changing the official name to *midori*, in 1973 they changed all the go lights to a blue-green colour – still within the definitions of green to satisfy the international traffic codes, but blue enough to fit the word *ao*. Definitely a case of language changing reality.

Studies on bilinguals have tried to find out if their view of the world is different, depending on which language they use. Russian-English bilinguals were asked to explain how the world works. When they spoke in Russian they expressed much more collectivist ideas and when they spoke English they spouted more individualistic values. Another study of Japanese women living in the United States asked them to complete sentences in Japanese and then in English, The first sentence was 'When my wishes conflict with my family's . . .' The Japanese response was 'it is a time of great unhappiness'. The English was 'I do what I want'. Another sentence, 'Real friends should . . .' was completed in Japanese with 'help each other'. In English it was 'be very frank'.

Languages around the world reflect different understandings of time and space. For example, in English, we use prepositions to express our idea of the past 'behind' us and the future 'ahead'. But some Aboriginal tribes use compass points rather than prepositions to describe both place and time. So time for them moves east to west, not forwards or backwards.

Professor Lera Boroditsky is a psycholinguist at California's Stanford University. She explains that in many languages there are words like *left* and *right* that are used to divide up space. 'In English we tend to divide our space relative to our bodies,' she says. 'If I turn through 180 degrees, the chair that was on my left is now on my right. In some languages words like *left* and *right* aren't used at all, and instead everything is expressed in some kind of geographical system. Sometimes it's north, south, east, west; sometimes it's relative to a hill that you live on, or a river that runs through your land. So everything will be up river, down river or across the river. In order to speak a language like that you have to stay oriented with respect to the landscape. You have to be able to say things like there's an ant on your south-west leg, but if you turn 180 degrees then it will be your north-east leg. So it's not on your body, it's really on the landscape.'

But does all this mean that it's the language which is shaping thought and perception, determining the way we perceive the world, or is it merely pragmatic? After all, it makes sense for a nomadic people to orientate themselves to their environment through compass points because they travel huge distances using landmarks. So what we do is shape our language to our environment – a bit like putting on a suit of clothes which we can adapt depending on the weather, occasion, environment, mood. You can tie yourself in knots over this debate – as many have done.

The question is, has the language developed the way it has because the culture has demanded that of the language? It's very hard to know which is the chicken and which is the egg. Here's Professor Boroditsky on the subject. 'Chicken and egg isn't the right way to think about it, because humans develop languages as tools for communicating and also as tools for thinking. And we develop the tools that we need, that we want, that we desire in our environment, to achieve the goals that we have. But then the tools that we create also give us cognitive abilities and allow us to do more. That's why we created them. And so the next generation of speakers really benefits from having those tools built into the language. And these forces mutually influence and reinforce each other.'

What is clear is that languages have always been evolving, often at remarkable speed, and when you look at where our own language, English, came from, you can see just how far it has travelled. And for that historical perspective we have to thank the detective work of two remarkable brothers.

Fairy Tales and PIE

The two brothers were both German professors, and their research into the relationship between similar words of different languages was an important contribution to the study of how language develops. But we know them better as story collectors, the purveyors of tales of wicked stepmothers and frog princes and sleeping princesses that permeate all of our childhoods. They were, of course, the Brothers Grimm.

The tale of this inseparable pair is a compelling one. Two brothers, just a year apart in age, who lived, studied and wrote together for almost their entire lives. Jacob and Wilhelm Grimm were born in the German town of Hanau, near Frankfurt, towards the end of the eighteenth century. They were the oldest surviving sons in a family of nine children – eight boys and one girl, three of whom died in infancy. Their father Philip was a lawyer and court official, and their early childhood seems to have been happy. Then, when Jacob was eleven, their father died. The family was forced to move from the countryside into a cramped townhouse, and the two boys were sent away to a school in the north. Jacob and Wilhelm went on to study law at university but after their mother died they both took positions as librarians to support their younger brothers and sister.

Both young men were passionate about German folklore and set out on a mission to preserve Germany's oral tradition and investigate how words change their sounds over time. As part of their research they collected fairy and folk tales from friends and guests, who often recounted stories told to them by their servants; from local peasants and villagers; and from published works from other languages and cultures. The story of Red Riding Hood is thought to have been recounted by Dortchen, a pharmacist's daughter and childhood friend whom Wilhelm later married.

They published their first collection of eighty-six German fairy stories, *Kinder- und Hausmärchen* (Tales of Children and the Home), in 1812. The volume included such classics as 'Cinderella', 'Snow White', 'Hansel and Gretel' and 'The Frog Prince'. The brothers clearly didn't intend the tales to be read by children or non-scholars. The first edition had a long introduction, reams of notes and no illustrations, and the stories themselves were certainly not child-friendly. These were earthy tales, full of cruelty and hunger and loss, of witches and trolls and ravaging wolves prowling through dark forests. Mothers, not just stepmothers, were evil; a wicked queen was forced to dance in red-hot shoes until she died; a servant was put into a nail-studded barrel and a witch baked alive; children were abducted by a pipe-playing rat catcher; even the Frog King was thrown against a wall to wake him up instead of being kissed. Not Walt Disney, then.

Wilhelm Carl Grimm and Jacob Ludwig Carl Grimm, noted for their work in linguistics as well as their famous fairy tales

The brothers' different personalities complemented each other. Bachelor Jacob – introverted, physically stronger – was the more academic of the two. He did much of the research for the folklore and the bulk of the work on the theories of language and grammar. Wilhelm was warmer, less physically strong – he had asthma and a weak heart – and he was more interested in literature; he edited and constantly revised the stories to give them their classic narrative style. The brothers continued to collect and publish hundreds of German fairy tales and legends – with limited financial rewards. Fortunately for them, Europe was experiencing a flowering of children's literature, and illustrated books of folk tales like 'Jack and the Beanstalk' were already hugely popular in England. Eventually, after the success of the first English translation of the Grimms' collection in 1823 (*German Popular Stories*, illustrated by George Cruickshank), the brothers selected fifty of their most popular tales, refined and softened many of them and republished the cheaper *Kleine Ausgabe* (Small Edition) in 1925. Their younger brother, Ludwig, illustrated it.

The Grimms' Fairy Tales, as the English-language version was called, were soon established as classic children's literature. The collection of tales became the most reprinted book in Germany after Luther's bible and has been translated into more than 160 languages. And what would Walt Disney, who kick-started his movie empire with *Snow White and the Seven Dwarfs*, have done without the Brothers Grimm?

As the brothers collected their folk tales they were able to investigate phonetic shifts in the German language which led them to formulate one of the earliest and most important linguistic laws, now known as Grimm's Law. It was largely Jacob Grimm's work. He traced the vowel shifts in various European and Ayran languages and found they were all linked to a common root. His work on linguistics, contained in the original German grammar book, *Deutsche Grammatik*, in 1819, was the first to employ a rigorous scientific methodology and could be said to have given linguistics – or philology as the Germans like to call it – a sound basis.

In 1838, the brothers began a monumental project: the *Deutsche Wörterbuch* (or simply *Der Grimm*) is to the German language what

The Grimms' fairy tales quickly became classics in children's literature

the *Oxford English Dictionary* is to English. The brothers planned to record information about the history and usage of every German word. They thought the task would take them ten years; in fact, Wilhelm had died by the time the first volume, *A–Biermolke*, was published, and when Jacob died in 1864, he was only at *F* for *Frucht* (fruit). An army of academics continued work on the dictionary, but it wasn't completed until 1961 – almost a century and thirty-two volumes later. It has around 350,000 entries and weighs 84 kilos. The dictionary is already being revised, and letters *A–F* are due for completion in 2012.

The brothers didn't live to see the fulfilment of another dream of theirs – the founding of the modern German nation-state in 1871. Jacob Grimm had written in his autobiography a few years earlier:

> *Nearly all my labours have been devoted, either directly or indirectly, to the investigation of our earlier language, poetry and laws. These studies may have appeared to many, and may still appear, useless; to me they have always seemed a noble and earnest task, definitely and inseparably connected with our common fatherland, and calculated to foster the love of it.*

With hindsight, Jacob's nationalistic tones – however romantic – are rather unsettling. I much prefer the words he used in his eulogy for his beloved brother Wilhelm. He called him simply *mein Märchenbruder*, my fairy-tale brother.

It's reckoned that there are more than 6,000 languages spoken around the world today. Most modern linguists don't think that these languages originated in just one place, but rather that a variety of them evolved independently among different groups. Current thinking is that almost all of them can be traced back to about ten main 'language families'. Chinese, for instance, derives from Sino-Tibetan; Swahili from Niger-Congo. But the largest of these families, the one that half the world's population can trace their language back to, is Proto-Indo-European.

Scholars in Europe first began to investigate the history of languages at the end of the eighteenth century. There was already clear evidence that a group of languages – French, Spanish, Italian and other Romance languages – were descended from Latin. What Jacob Grimm, in his *Deutsche Grammatik*, and a handful of others did was to trace back almost all the languages of Europe, together with some from western Asia and northern India, to one single ancestor language – Proto-Indo-European (PIE).

Grimm's Law, as it became known, showed how the consonants of different Indo-European languages relate to each other. For example, there is a regular relationship between words beginning with *p* in Sanskrit, Latin or Greek and *f* in Germanic languages (including English). So, *pater* in Latin becomes *father* in English; or *pada*, *ped* and *pus* in Sanskrit, Latin and Greek become *foot*, *voet*, *Fuss*, *fotur* and *fod* in English, Dutch, German, Icelandic and Danish. Other examples show the relationship of *k* to *h* in non-Germanic and Germanic languages: *kyon*, *canis* and *ci* (dog) in Greek, Latin and Welsh become *hound* and *Hund* in English and German. And there's also a link from *g* to *k*, for example, with *gelandros* and *gelu*, meaning *cold* in Greek and Latin, becoming *cold*, *kalt* and *kall* in English, German and Swedish.

We think this single language was spoken more than 5,000 years ago in the Steppes of southern Russia. As tribes migrated through Europe and Asia, PIE split into a number of dialects, and these in time developed into separate languages. It wasn't the first language of man; it's simply the oldest we have evidence of. We don't know what PIE would have sounded like, and there are no written records. All we can do is try to inch our way tantalizingly nearer by looking for clues in the oldest written records we've found so far. For example, scholars have worked on prose from the Old Indic literature of ancient India, which was passed orally from generation to generation. One piece of prose – *The King and the God* – is thought to be the oldest sample of literature in any Indo-European language. By working backwards a few millennia, this is how some scholars think a PIE storyteller 5,000 years ago might have sounded: 'To réecs éhest. So nputlos éhest. So réecs súhnum éwelt. Só tóso cceutérm prcscet: "Súhnus moi jnhyotaam!"' ('Once there was a king. He was childless. The king wanted a son. He asked his priest: "May a son be born to me!"').

73

The floor of the General Assembly of the United Nations in New York looks a bit like a reincarnation of the Tower of Babel. Hundreds of voices talking in different tongues, evidence that mankind's initial proto-languages split into thousands more. In fact, there are just six official languages in the United Nations – English, French, Spanish, Chinese, Russian and Arabic; de facto, all 192 member states have to speak one of these to be heard. It is rather ironic that the United Nations, the defender of the rights and sovereignty of all its member states, has to use old-fashioned language imperialism to make itself understood. Perhaps the world would be better off if everybody spoke just one language.

'Oh, no,' insists Zaha Bustema, an Arab-English simultaneous translator. 'No, no, no, no!' Her passionate response isn't simply because it would put her out of a job. 'There is a beauty of languages. They are living entities. Each and every language has its own music, its own imagery, its own way of expressing the sentiments and the nature of the people. It would be a loss if that language did not exist. Oh, I'm very much in favour of the Tower of Babel.'

That Tower of Babel is increasingly under threat as the homogenization of languages seems to be pushing us back to the days of a few proto-languages; and that, as we shall find out in the next chapter, spells the death of many of our micro-languages, each one a repository of knowledge, a definer of culture and identity, exquisitely structured and vastly complex.

The United Nations headquarters in New York, 2010

CHAPTER 2

Identity

There are around 195 countries in the world, and for many the name of the country refers to the language they speak – English for England, German for Germany, Chinese for China. Of the more than 6,000 languages on our planet, some are spoken by only a score of people living in a valley in Papua New Guinea, others, like Chinese, by over a billion. But scratch beneath the surface and, of course, even the seeming homogeneity of Chinese is an illusion; Chinese has in fact got seven distinct languages with many more dialects, although unified by one writing system. Then there are another 292 distinct ethnic languages within the People's Republic of China.

So what has our language, our mother tongue, *langue maternelle*, *Muttersprache*, *bahasa ibu*, to do with our identity? What is certain is that language has been used to unify and oppress, liberate and imprison. It is part and parcel of conquest – be it Latin in the Roman Empire, Spanish in Latin America, Norman in Britain or English in North America. But lest we think it is all about Western imperial powers, over 600 years ago the Incas did it just as effectively, imposing their language, Quechua, on hundreds of tribes on the *altiplano* that they assimilated during the creation of their empire based around Cuzco in present-day Peru.

Language has always been a powerful weapon used to dominate and subjugate; far too often this has meant the proscription and eventual demise of the smaller languages. 'Learn well the language of the whites. Do not rely on our language, there's no value there. One's future well-being is dependent on mastering the language of the foreign people.' So spoke a Hawaiian to her granddaughter in 1896, and it is a tale that could be told all over the world. Little surprise, then, that we are losing languages at the rate of one every fortnight. By the end of the century it's estimated there will be only 900 languages left in the world.

It matters because, more than an ID document or a passport, what defines us is our language. Go anywhere in the world and you will find examples of how linguicide, the death of languages, is having a devastating effect on identity. But there are also a few places where minority languages are fighting back.

Amongst the Turkana

Africa, the last victim of the colonial period, is a witches' brew of languages. There are an estimated 1,500 different languages spoken in the fifty-two countries on the continent and for the fragile post-colonial nation-states this Babel is seen as an impediment to forging national identity and development.

The area round Lake Turkana – the Jade Sea or Lake Rudolph as it used to be called in school geography books – is the sort of habitat where Lucy, our early hominid progenitor, scrabbled about, hunting kudu and eking out a subsistence. The lake, in best post-colonial fashion, is now named after the predominant tribe of the area, the Turkana. They are a feared and uncontrollable lot and, for the British trying to 'pacify' this northern area of Kenya, as prickly as the acacia trees that define the landscape.

The Turkana tribe number some half a million and live in one of the most inhospitable places on earth: a semi-arid desert that stretches across northern Kenya and southern Sudan. They are a fierce and fearsome people who have defied colonization by both whites and blacks for over a hundred years but are now, through education, being brought into the fold of modern Kenya. But a consequence of this development is they risk losing not only their language but also their very identity.

The Turkana live for their livestock – oxen, camels, goats and sheep, in that order of importance – and will do anything to protect them. As a result they have become toughened warriors, much feared by their neighbours. They are monotheistic in the sense that they have a creator, Akuj, who is the bringer of rain, which to the Turkana is synonymous with life. Their culture is predicated on raiding the livestock of their neighbouring tribes, predominantly the Toposa over the border in Sudan and the Karamajong, who are spread out in an area spanning Kenya and Uganda. They love fighting as much as they do their oxen. All the young men, rather unnervingly, carry Kalashnikovs. Wealth is counted in livestock and not much else for the men whilst the women are adorned with an extravagant number of beaded necklaces – so many that their necks are never seen. A rich woman will wear a good 5 kilos of beads, and they are never taken off.

Turkana women wear up to 5 kilos of beads around their necks that they never take off

Being nomadic, the Turkana did not take kindly to the imperial powers' ludicrous method of creating borders. Their land was split between Sudan, Uganda and Kenya, so the borders were irrational to them. The one with Sudan seemed the work of a particularly indolent cartographer who must have got bored and simply drawn a straight line. Turkanaland was eventually integrated into Kenya, albeit cut off from the rest of the country by the Closed Districts Ordinance, which prevented all but a few officials from entering the area. It was only in 1964, after independence, that President Jomo Kenyatta repealed the Ordinance, and it was deemed safe enough to let people travel there. Few did. The Turkana were left alone for most of the twentieth century, and the changes in their lifestyle were minimal. Their survival continued to rely heavily on the rains, and the excitement of cattle raiding carried on as usual. Isolated from the political and economic life of the rest of the country, they felt little pressure to adapt to the changing face of postcolonial Kenya. Their identity was unchallenged.

As always in Africa the sheer number of tribes within each nation-state has created linguistic problems. The Turkana language is just one of 69 in Kenya. Tanzania has 128, Congo 242, Nigeria more than 500. Tribal allegiances are still the most powerful forces within these countries, and mutual incomprehension impedes social and political cohesion. You only have to look at the history of civil wars from Biafra to Rwanda to see the tragic consequences. The answer in Kenya, as indeed in much of east Africa, has been to adopt Swahili as a lingua franca, so that at least there's one language which everyone can speak and understand.

Swahili (from the Arabic *sahil*, meaning 'coastal dwellers') is an odd creation. Originally a language for trade, it mixed local Bantu dialects along the east African coast with elements of Arabic, Portuguese, French, German and English. It was barely a written language but after independence it was the natural choice to become the language of the new states. Tanzania, under its first president, the visionary socialist Julius Nyerere, adopted it vigorously, hoping that it would eliminate tribal differences and unite the country. Jomo Kenyatta in next-door Kenya similarly adopted it as the lingua franca and language of instruction in schools, albeit with English still being widely taught and spoken.

English- and Swahili-language newspapers on sale in Africa

For the Turkana, Swahili had some uses for trading but not much else. Schools were non-existent until very recently, and it was the work of the Catholic missionaries in the 1960s that introduced a written script into a hitherto entirely oral language

The market in Loki is a bustle of Turkana, who've walked in from the desert to shop at the Arab and Somali stalls selling everything from mobile phones to beads and foodstuffs. All the trade with non-Turkana shopkeepers (about 95 per cent) is being conducted in Swahili. It's not particularly elegant Swahili but it works. Speaking Swahili is just a practical necessity. But in Loki's government-run school, it's evident that Turkana is not being taught at all. Instruction is either in Swahili or English for some subjects. So the young Turkana who do go to school, albeit still a small percentage of the population, are in danger of losing their mother tongue – the pressure to conform and speak either Swahili or English is just too great.

For those who remain in their traditional *manyattas* (villages) the chances to do other than be a pastoral nomad are negligible. For the older generation this is not really a dilemma – they have known no other way of life – but for their children, if they don't embrace the possibilities of education and the chance to become involved in the wider Kenyan society, they are destined to remain on the periphery. The upside to this is they will retain their language culture and identity, but how long they can withstand the somewhat dubious temptations of 'progress' is a moot point.

One of the key factors in preserving the language has been the creation of a script. Thanks mainly to the work of Catholic missionaries like Father Tom, an Irish priest who has been working among the Turkana for nigh on forty years, a standard Turkana orthography has been adopted. It is based on the Roman alphabet and is taught at the mission primary schools. Under the shade of a tree a young teacher gathers her children and, with nothing more than a worn-out blackboard, chalk and the efficacy of using the desert sand as a slate substitute, the children practise writing Turkana. The eagerness with which the young Turkana sing out the vowels and write them down is an inspiring sight. However, unless their language is adopted by the state system it seems it is never going to provide a realistic alternative to Swahili.

Turkana children learn to write by drawing in the sand

The *manyatta* is a world apart from the relative sophistication of Loki, a half day's walk away. To the west you can see the outline of the Kidopo mountains, which form the border with Uganda and were made famous in the 1970s by director Peter Brook's play based on the Ik, the tribe that live there. Five miles north across the flat scrub desert lies the border with Sudan (or whatever the new state of Southern Sudan will be called). The fact that Southern Sudan has effectively abrogated the colonial borders and seceded from the very idea of Sudan is indicative of the looseness of nation-state identity in parts of Africa. Tribalism rules. This is an unforgiving landscape, where tribal and clan loyalties are paramount, and the only way to assert security and independence from larger forces is to be united. Language is, as ever, the key to this sense of belonging.

The *manyatta* itself is almost invisible from a distance: a scattering of thorn-bush enclosures for the livestock and domed huts made of bendy boughs and stretched animal skins that seem to have grown out of the earth, which of course they have. The giveaway signs are the bleets, baahs and moos of a few hundred quadrupeds and the waft of animal urine from the enclosures.

The young men here have no intention of giving up their traditional way of life, which means basically cattle raiding. The notion that there is anything more important than Turkanaland and the Turkana language and customs is clearly unthinkable. Cedrac Ekitela is a strikingly handsome warrior who seems to have some authority even in this fiercely consensual society. He has the additional status of being the proud owner of a shiny blue mobile phone, and to add to his skills, speaks passable Swahili with a smattering of English. He explains that to express concepts such as Akuj, their supreme god and bringer of rain, or to sing the *naleyo*, which honours their *emong*, a special ox who becomes like an avatar to the warrior, would be impossible in any other language. Turkana is unusually rich in mythology and songs, and it is this inner part of the Turkana which no other language can reach. Anyone watching them perform these ritual songs and dances would see that their language is their culture.

Early explorers and British administrators must have felt intimidated when confronted by these uncompromising warrior people.

It is this that will undoubtedly help them preserve their language and traditions more than anything else; and their half a million plus numbers – relatively substantial by Kenyan standards – will ensure they have enough weight to protect themselves not only from the neighbouring tribes but also political intrusions from Nairobi.

Their unique voice and vision of the world should not be lost.

Turkana traditions will hopefully be around for years to come

Turkana Origin of Stars Myth

A taboo is broken. Akuj, the bringer of rain, is angry and covers the world in darkness with his cloak. The Turkana warriors all get together and with their spears lift up Akuj's cloak. The spears make tiny tears in the fabric and those holes are the stars that shine at night.

The Akha: the Politics of Language

The Turkana have a good chance of speaking their language in ninety years' time, but halfway across the world in Thailand a similar sort of story is unfolding which may not have a happy ending.

Chiang Rai is the northernmost province of Thailand, and its eponymous capital is only 60 kilometres from the borders with Burma (Myanmar) and Laos. A great sprawl of shops offering everything from new Volvos to durian fruit line the road north as far as the newly created Mae Fah Lung University, named in honour of the much-loved mother of the present King of Thailand. Beyond the campus area, the shops give way to padi and tobacco fields, and the road begins to climb up through the hills. To an outsider the villages along the road all look much the same, but there are some telltale signs that point to the complex ethnic and historical forces at play here. The occasional flash of a traditional headdress or colourful

Traditional bamboo houses line the hills

fabric amid the sea of ubiquitous jeans and T-shirts mark out the Hmong from the Lisu, the Lahu, the Shan or the Wa. These are the main tribes that live in these highlands, which extend north up to Yunnan in China, east across Laos as far as the Annamite range in Vietnam and west to the Salween river in Burma. They all have their own languages and customs, and none of them owes any deep-rooted allegiance to the parent state of which they, by accident of history, find themselves part. It's a rich ethnographic smorgasbord. The mountains offer both protection and isolation, allowing fascinating and discrete cultures, and of course languages, to develop.

The road finally reaches Kuay Huak Paso, a straggly line of bamboo and concrete houses that line the ridge that separates Thailand from Burma. It is home to one of the many hill tribes of Thailand, the Akha.

Aju is an Akha but he speaks good English, having studied at language school in Oxford. He stayed with the poet James Fenton for a couple of years before returning to Thailand to fight for Akha rights. The Akha are one of the poorest and least developed hill tribes in Thailand, looked down upon by the Thais, and often refused IDs, which means they have little or no access to health and welfare programmes. Aju set up an Akha institute to lobby for these rights and has begun to record the oral traditions of his tribe before they die out. He wants the government to acknowledge that the Akha language, along with other minority hill-tribe languages, has a right to exist and should be helped to survive.

Today, Aju is visiting the village blacksmith's, where a very rare and important event is about to start, a ritual that even Aju, video recorder of his tribe's culture, has never witnessed. As the blacksmith begins to fashion a blade from a solid rectangular lump of steel, ably helped by his wiry assistant working a most efficient and excellent bamboo bellows, Aju explains the significance. The elders of the village are about to embark on a three-day verbal marathon, relating the entire Akha genealogy to one of the younger men. Traditionally the Akha have had a purely oral culture – preserving family histories, children's songs, games and stories by remembering and repeating them from generation to generation. Aju can recite fifty-three generations of his family tree. Not bad. Most of us would be hard pushed to remember four. We have the advantage of history books, diaries, literature to remember both our personal history and that of our culture. For the Akha to lose their language would effectively erase a thousand years of history and render them mute.

Aju isn't convinced the Akha language will survive

Although Akha does not have an established writing system, various missionary sects have created nine alternatives. There's a Catholic one, a Baptist one, a Lutheran one, a Methodist one, all tinged with the underlying premise that the Akha should convert to Christianity. Aju doesn't approve of missionaries. He explains that the Akha have not yet decided on a system for writing their language, but it will be a new one, created for and by Akha and based on the Roman alphabet with some special ways to signify the tonal nature of their language.

Oddly, the Akha, for so long in contact with the dominant cultures that do have writing, have never had a system for recording, be it alphabet or ideogram. Aju says, 'In the old days, I don't know how many years ago, they say Un Ma, the first spirit, gave an alphabet to all people and, because he'd run out of paper, he presented the

Akha tribe with a water buffalo's hide on which the Akha letters were written. The Akha took the skin home and it was fine. But then one day there was a great storm, and it got wet. So they took the skin near the fire to dry it out. But the smell got so nice that they couldn't resist eating it, and since then the Akha have had no alphabet. We ate it!'

The blacksmith hammers away, and before long a blade that will be used to kill an ox and finish the ceremony takes shape. It's all very relaxed, and a lot of rice wine is consumed as the elders begin their recitation, which is nothing less than the history of the tribe. The sixty generations that the initiate will have to learn tell of the Akha's migration through southern Yunnan, Burma and Laos to their present place high on this saddleback in the Golden Triangle. It seems not dissimilar to those biblical passages where begat is a much overused verb.

It's old men doing their old men's stuff (no women take part). Aju confesses his own daughters are more interested in their Facebook friends, Twitter and watching YouTube. And therein lies the rub. To communicate with the larger world the Akha must learn either Thai or English.

The village primary school is new, a well-built block of five classrooms, adorned with various pictures of the King of Thailand and more Thai flags than the national airline. A football pitch takes centre stage with the dramatic hills behind that define the border with Burma. Money has obviously been spent on it. The village is only 5 miles from the border with Burma, and the Thai government is keen to instil a sense of national identity in these hill tribes whose allegiance is, at best, fluid. The key to this is language. Thai is the compulsory medium of all instruction, and even if a universal Akha script existed it probably would not be taught even as a second language, English being given preference. But Aju is adamant that the Akha, along with the million plus other hill-tribe people, need to have their written language to maintain their identity and at least give them a chance to maintain their rich culture and heritage. Linguicide leads to cultural genocide within a generation.

Aju is not convinced that the Akha language will survive for many more generations, but he is determined to do his best to preserve it.

Thai and English are taught in Akha schools

He explains how he recently went to a meeting of hill-tribe people in Yunnan in southern China, and they agreed on the style of the new Akha script. But it may be too late. Aju's daughters don't even speak Akha at home that much, preferring to speak Thai. Wistfully Aju laments: 'At my home I speak Akha language, my wife and I speak Akha to them, but they reply to me all in Thai. But they understand Akha. Most of the Akha, especially the younger generation who moved to the city, they don't show to the outsider that they are Akha because we used to be looked down on by the majority of the Thai people, and they said the Akha people were dirty and lived in the forest, and didn't have a proper language. And this memory transfers to the kids; that's why the kids want to hide it. They don't want to be Akha.' Like so many minorities, they feel slightly ashamed of their own language with its perceived 'primitiveness'. So they do their best, like all children, to fit into the dominant culture.

To lose a language is to lose a culture and its history. Homogenization of both language and culture seems inevitable in this part of the world, but, while it might seem cleaner, easier and safer, it diminishes us. It may not seem as bad as the extinction of a whole species, but it is as great a loss.

Irish: Famine and Revival

The west coast of Ireland on a bitterly cold December morning is one of the most unusual – certainly the most beautiful – places to play a round of winter golf. The Connemara Isles golf club is on the westernmost tip of Europe and Connemara in County Galway is part of what is called the 'Gaeltacht', one of the central areas for the speaking of the ancient language of Ireland – Irish. 'Everybody here thinks in Irish,' one of the golfers explains. 'You can get through your lifetime here without speaking English.'

The thatched club house is a reminder of a catastrophic event which helped to shape the story of the Irish language. It was built by the great grandfather of the current owners in 1850. He was one of a handful of survivors from the wreckage of the famine ship *Brig St John*, which foundered on rocks near Boston Harbour in October 1849.

The *Brig St John* was a so-called coffin ship, one of the many ill-equipped, overcrowded vessels which carried hundreds of thousands of Irish emigrants across the Atlantic to the east coast of America between 1845 and 1851. This was the time of *An Gorta Mór* – the Great Hunger – better known outside Ireland as the Potato Famine. Around a million people – one in eight of Ireland's population – are thought to have died from starvation or disease; another million emigrated from Ireland to the States, Canada, Australia, New Zealand and Great Britain. The Irish potato famine was probably the worst human disaster in nineteenth-century Europe and a watershed for Ireland, politically and culturally; it sounded the death knell for the Irish language.

It's difficult to get your head round just how catastrophic *An Gorta Mór* was. There aren't many written accounts from survivors; a bit like people who lived through the horrors of the Battle of the Somme or the Nazi concentration camps, they simply didn't or couldn't talk about it. There was extensive newspaper coverage at the time, however. One of the worst-affected areas was County Cork in southwest Ireland; the town of Skibbereen – where one in four of the population is thought to have died – became the focus for press and outsider interest. The *Illustrated London News* hired its own special correspondent, illustrator James Mahony, to report from Skibbereen. And, in a world before printed photographs, the illustrations accompanying Mahony's graphic two-part series became synonymous with the famine.

> *We next reached Skibbereen . . . and there I saw the dying, the living, and the dead, lying indiscriminately upon the same floor, without anything between them and the cold earth, save a few miserable rags upon them. To point to any particular house as a proof of this would be a waste of time, as all were in the same state; and, not a single house out of 500 could boast of being free from death and fever, though several could be pointed out with the dead lying close to the living for the space of three or four, even six days, without any effort being made to remove the bodies to a last resting place.*

People in Skibbereen died in such huge numbers and so fast that there weren't enough people alive to bury them. Pits were dug near the ruins of the old Abbeystrewery Franciscan friary, and up to 10,000 coffin-less bodies were buried in the mass grave.

Heated debate is still in full swing over whether the Great Hunger was the result of a natural catastrophe worsened by the British government's inaction or a form of genocide. It's a historically and politically charged issue. What is without question is that the famine was a huge turning point for Ireland. Indeed, historians talk about Irish history in terms of the pre-famine and post-famine years.

The famine hit rural areas hardest and in particular the Irish-speaking areas of western Ireland. In 1835 there were an estimated 4 million native Irish speakers. This number had fallen to 2 million by 1851 and to 680,000 by 1891. Those who hadn't been killed by the famine emigrated or chose to speak English in preparation for finding employment abroad or in the cities.

We need to delve a bit further back into history here. The Irish language grew out of the Proto-Celtic language which was spoken all over western mainland Europe from around the fifth century BC.

BOY AND GIRL AT CAHERA.

James Mahony's drawings recorded the desperation of the Irish during An Gorta Mór

Groups of Celtic-speakers migrated to Ireland around the fourth century BC, where a form of Celtic known as Goidelic (or Gaelic) developed; fragments of primitive Irish are known from inscriptions in the Ogham alphabet, etched into stone monuments in southern Ireland around the fourth century AD. After Ireland's conversion to Christianity and the introduction of the Roman alphabet in the sixth century, the Golden Age of Old Irish began. This evolved into Middle Irish, which then spread through Ireland and into Scotland and the Isle of Man. Modern Irish, with the assimilation of Old Irish, Norse, Norman and Old English, emerged around the twelfth century. In Scotland it became Scottish Gaelic; in the Isle of Man, the Manx language.

Another group of early Celtic-speakers migrated to the British mainland around the same time as Goidelic was developing in Ireland. The language of this group, which settled in southern England and Wales, grew into a type of Celtic known as Brythonic (or British); this provided the basis for Cornish, Welsh and the now extinct Cumbric; and a movement of some of these settlers from southern England across the Channel to Brittany around AD 600 led to the development of Breton there. As Old and Middle English developed from the languages of the invading Anglo-Saxons and then the Normans, the Brythonic Celtic languages were driven into Wales and the south-west.

Thus we have Modern Irish being spoken by almost the entire population of Ireland in the twelfth century. Over the next few centuries there were attempts by the Normans and then by the English to control the territories of the Irish lords, and by the beginning of the sixteenth century there was an English-controlled territory on the east coast, with Dublin as its centre, known as the Pale. Here the English language predominated. Outside this area – beyond the Pale – were the Gaelic Irish, who maintained their own language and laws. The English called them 'His Majesty's Irish enemies'.

The assault on the Irish and their language began in earnest in the mid sixteenth century when Henry VIII and subsequent monarchs confiscated land and replaced the Irish Catholic land-owners with settlers from England and Scotland. The Plantations,

Jonathan Swift recommended the Irish sell their children for food

as it came to be known, was a ruthless policy of colonization. The Crown took over large swathes of land in the south-west province of Munster and in the northern province of Ulster – the most remote and Gaelic of all the Irish territories. As the Irish tongue shrank westwards and the Irish-speaking ruling classes disappeared, Gaelic Irish culture was sidelined, and a new English-speaking elite dominated the politics of the country. English was the language of parliament, law and trade and, after the introduction of the National Schools system in the 1830s, the language of the classroom as well.

It's easy to forget that some of the English language's most famous writers were from this Anglo-Irish elite. The satirist and essayist Jonathan Swift was born in Dublin in 1667. His *Modest Proposal (for Preventing the Children of Poor People in Ireland Being a Burden to Their Parents or Country, and for Making Them Beneficial to the Public)* recommends that Ireland's poor escape their poverty by selling their children as food to the rich: 'I have been assured by a very knowing American of my acquaintance in London, that a young healthy child well nursed is at a year old a most delicious, nourishing, and wholesome food . . .' Oscar Wilde was born in Dublin in 1854, at the end of the Great Hunger. As was George Bernard Shaw, two years later.

The Great Hunger virtually killed off the already struggling native language, but it hardened resentment towards the British and became a rallying point for nationalist movements. The spearheading of a Gaelic revival came from the very section of Ireland's society which nationalists blamed for killing off the language in the first place – English-speaking Protestants. Douglas Hyde, academic, passionate supporter of the Irish language and future President of Ireland, was the Anglo-Irish son of a Church of Ireland rector. He taught himself to speak and write Irish, joined the Society for the Preservation of the Irish Language and penned dozens of Irish verses. In 1889, he wrote: 'If we allow one of the finest and richest languages in Europe, which, fifty years ago, was spoken by nearly four millions of Irishmen, to die without a struggle, it will be an everlasting disgrace and a blighting stigma upon our nationality.'

Hyde helped set up the Irish National Literary Society with W. B. Yeats in 1892 and at one of its first meetings concluded a lecture entitled 'The Necessity for De-anglicising the Irish People' with a literary call to arms:

> *I would earnestly appeal to everyone, whether Unionist or Nationalist . . . to set his face against the constant running to England for our books, literature, music, games, fashions and ideas. I appeal to everyone . . . to help the Irish race to develop in future upon Irish lines . . . because upon Irish lines alone can the Irish race once more become what it was of yore – one of the most original, artistic, literary, and charming people of Europe.*

In 1893, Hyde helped to found the Gaelic League (Conradh na Gaeilge). Its aims were to preserve spoken Irish, restore it as the country's first language and encourage the revival and creation of literature in Gaelic.

The League faced an uphill battle to convince Ireland's dwindling number of native speakers that there was any point in saving the language. It was helped by the growing strength of the nationalist movement, which saw the speaking of Irish as a statement of national pride and an act of defiance against the English. Although the League was non-political, it attracted future nationalist leaders. Patrick Pearse, one of the Easter Uprising leaders, joined when he was sixteen and became editor of its newspaper. The Irish Republic's founding father, Eamonn de Valera, met his wife Sinead at the League, where she was teaching Irish. The Irish-speaking parents of Michael Collins, the Sinn Féin leader, had insisted on speaking English to their children so they would 'get on' and find good jobs in an English-speaking world. Michael Collins was passionate about the League.

> *The Gaelic League restored the language to its place in the reverence of the people. It revived Gaelic culture. While being non-political, it was by its very nature intensely national. Within its folds were nurtured the men*

and women who were to win for Ireland the power to achieve national freedom. Irish history will recognise in the birth of the Gaelic League in 1893 the most important event of the nineteenth century. I may go further and say, not only the nineteenth century, but in the whole history of our nation. (Michael Collins, The Path To Freedom)

The League proved immensely popular in English-speaking areas of Ireland and by 1908 it had around 600 branches. The non-political stance of Hyde became increasingly unpopular and in 1915 he resigned the League's presidency after its members committed it to 'a free, Gaelic-speaking Ireland' at the annual conference. Many of those members participated in the Easter Rising the following year and helped set up the Irish Free State a few years later.

Hand in hand with the revitalization of the Irish language at the end of the nineteenth century was a literary revival which saw the flourishing of some of Ireland's greatest writers. Sitting at the centre of this extraordinary blossoming, in her black widow's weeds, was the figure of Lady Gregory, who, along with her friend William Butler Yeats, was the driving force of the revival.

Described by George Bernard Shaw as 'the greatest living Irish-woman', she was born Augusta Persse in 1852, the youngest daughter of a Unionist landowning family in Galway. At the age of twenty-seven, the plain-faced Augusta climbed to the top of the social ladder when she married widower Sir William Gregory, a retired Governor of Ceylon, who owned neighbouring Coole Park estate. He was thirty-five years older than his new wife. The couple spent much of the 1880s travelling or in London, where their house was filled with literary luminaries of the day – Browning, Tennyson, Henry James.

When her husband died in 1892, Augusta donned the widow's weeds which she wore until her own death forty years later and retreated to Galway, to the family house at Coole Park. Widowhood seemed to inspire her. She learned Irish and began collecting local stories and translating them into the Anglo-Irish dialect of her local area, Kiltartan.

The playwright Edward Martyn – later to become the first president of Sinn Féin – was a neighbour, and it was during one of her visits to his home at Tullira Castle that she met W. B. Yeats. This was the beginning of a lifelong friendship between the two. The redoubtable widow tried to teach the young writer Irish but with little success. Yeats is reputed to have tried and failed on thirteen separate occasions to learn to speak the language.

Lady Gregory's home at Coole Park became a summer retreat for Yeats and other writers of this Irish Revival. There's a giant copper beech still standing in the grounds of Coole Park known as the Autograph Tree. Carved into its trunk are the initials of its summer visitors, amongst them Yeats, J. M. Synge, Sean O'Casey, George Russell (AE) and George Bernard Shaw. Out of these long summer sojourns came the idea for the Irish Literary Theatre, which Lady Gregory, Yeats and Martyn set up in 1899. Its aims, according to its manifesto, were: 'to build up a Celtic and Irish school of dramatic literature . . . to bring upon the stage the deeper thoughts and emotions of Ireland'.

The project collapsed due to lack of funds, but the Irish National Theatre Society was set up soon after, and out of this came Dublin's famous Abbey Theatre – seen by many as the cornerstone of the Irish Revival. The Abbey was immediately popular with audiences, but first and foremost it was a writers' theatre, a hugely influential nursery for playwrights and actors.

George Bernard Shaw described Lady Gregory as 'the charwoman' of the Abbey; she turned her hand to anything and everything. She wrote over forty plays, directed, acted and raised funds. When crowds rioted at the opening of Synge's controversial *The Playboy of the Western World* in 1907, one observer described Lady Gregory as 'as calm and collected as Queen Victoria about to open a charity bazaar'.

Lady Gregory died in her beloved Coole Parke home in 1932, in a newly independent Ireland. Her influence on the revival of an Irish literature was huge; her effect on the revitalizing of the Irish language less so. The numbers of Irish-speakers in the census of 1926 had fallen to just over 500,000, the lowest level ever.

Lady Gregory was one of the few members of the Abbey Theatre

Lady Isabella Augusta Gregory, a founder of the Abbey Theatre in Dublin

The Autograph Tree, Coole Park

who spoke Irish well; most of the work to come out of the literary revival was written in English, or rather in Hiberno-English, the version of English spoken in Ireland. But this was an English never seen before, one word put after another in the service of a good sentence with more skill and artistry than had almost ever been seen before. The influence of the Irish language on the writing of Yeats and Joyce and Beckett was fundamental. Here's Declan Kiberd, Professor of Hiberno-Anglo Literature at University College Dublin, on the subject.

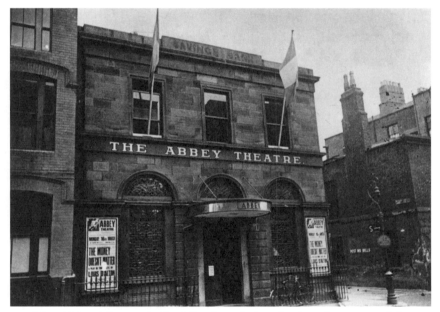

The Abbey Theatre in Dublin, founded to further the cause of the Irish Cultural Revival

'You have to remember that Irish as a language was spoken up till the 1840s and 50s by many, perhaps the majority of Irish people. So it's a backdrop to all this. People made the transition from Irish to English so rapidly that many were thinking in Irish while using English words and they taught themselves English out of books. So they acquired strange pronunciations of their own, they re-routed

the genius of the English language down new circuits, and I believe
this is one of the reasons for the explosion of writing at the beginning
of the twentieth century in Ireland. You can explain this in a very
concrete way. You know in Standard English, if you want to empha-
size a word you do it tonally: "Are *you* going to town tomorrow?"
or "Are you going to town *tomorrow*?" But in Irish you bring the
key word forward in the sentence, "Is it you that's going to town
tomorrow?" "Is it tomorrow that you're going to town?" And that
can sound quite poetic to an ear that's trained in Standard English
because it's a deviation. If you map this on to a wider attitude, to the
rules of literature, people like Joyce and Yeats and Beckett, they broke
so many of the rules with abandon because they didn't have any
superstitious investment in English literary traditions, they were in
the end at an angle to them. So the freedom with language becomes
a freedom with form in general.

'Hiberno-English has other grammatical idioms which are very
similar to Irish. For example, Irish has no pluperfect tense and uses
the phrase *tar eis*, which means "after". Hiberno-English imitates
this and, instead of saying "I have just done", it says "I was after
doing". "Why did you hit him?" "He was after punching me."

'Irish doesn't have words for yes and no and instead repeats the
verb in question. Hiberno-English uses yes and no much less frequently
than Standard English. "Is it cold outside?" "It is." "Are you coming
in?" "I am."'

Perhaps it was because Ireland's writers didn't have Milton or
Dryden or Tennyson breathing down their necks that they were able
to draw on the Irish language, or at least the memory of that language
– even though they couldn't speak it themselves.

Politics, so often the force which kills off a language, became the
Irish language's best friend. The Irish Free State established in 1922
– later to become the Republic of Ireland – made the restoration of
the Irish language part of government policy. The new constitution
declared Irish to be the national language; it was made compulsory
at school and speaking it was a prerequisite for any job in the civil
service. Gaeltacht areas – the native Irish-speaking areas – were officially recognized. These Gaeltachts are mostly in the remote areas of
the west of the country, in Donegal, Mayo, Galway, Kerry and Cork.

But the measures came too late – and the government's efforts were seen by many to be too heavy-handed. The teaching of the Irish language in schools tended to be dull, the learning difficult. Irish lessons came to be dreaded by schoolchildren in the same way as many are in terror of Latin – a dead language, grammatically difficult, removed from everyday life and to be dropped the minute one leaves school.

Today, most people in Ireland, including those in Gaeltacht areas, no longer speak Irish as a first language. The most recent figures suggest that only a quarter of households in the Gaeltacht areas are fluent Irish-speakers. Irish may be the first official language of the Republic but it's certainly not spoken by the majority of its people.

Attempts to keep Irish alive continue. Raidió na Gaeltachta (Gaeltacht Radio) and Teilifís na Gaeilge (Irish-language Television, or TG4) have had limited success, but the jewel in the crown is an award-winning TV soap in Irish called *Ros na Rún*. The entire crew from first AD to best boys, gaffers and cameramen all speak to each other in fluent Irish. They provide a useful sounding board for where people think the Irish language is now. The producer, Hugh Farley,

Stephen appearing in Irish-language soap Ros na Rún

Posters on a wall in Dublin promoting Irish

predicts that Irish is always going to be a second language for the majority of Irish people. He does, however, note a new 'coolness' about Irish.

'I know that if you were to speak to some of the cast here who grew up in the area they'd say that within a generation or so they can remember going into Galway City and being laughed at because they were speaking the Irish language. It was held in such low status. I think that has changed in recent years, particularly in Dublin because of the growth of Gael schools. So actually the great and the good, the bourgeoisie, wanted to send their kids to Irish-language school because they believed they were going to get a better education through the Irish language.'

The future of the language may indeed rely on the urban communities of Irish-speakers who are sending their children to Irish-language *gaelscoileanna* in droves. Nearly 10 per cent of all schoolchildren attend these schools, the fastest-growing sector of education in the country. All of the standard curriculum lessons are taught in Irish – apart from English, of course.

The pupils at one of these schools have interesting ideas about when it is that they use English and when Irish. English, for instance, is the language of the internet and definitely of texting. The grammar of written Irish is fiendishly complicated, so book-reading for pleasure is usually in English as well. No Gaelic Harry Potters here. But Irish is the language of the playground and out shopping with friends and chatting on the phone. It's the language of everyday life. Will they keep on speaking Irish, or is the language in terminal decline? Hugh Farley is unsure.

'I think that we have as a nation a very ambivalent attitude to the language . . . The Irish feel kind of culturally embarrassed that they are not good at the language and they are slightly resentful that they can't speak it fluently and therefore they try to discard it . . . I think that the trouble is we're caught in the double bind. We have people within the Gaeltacht areas [where] the younger generation are speaking it less than ever before and the number of people who are learning Irish and using it outside of the Gaeltacht is increasing. So we may be in a situation in twenty years' time where there are more people who are non-native speakers of the language who are keeping the language alive rather than people who live indigenously and speak it natively.'

Nipperlings and Waterstuff – the English Story

Think about these pairs of words for a minute: *lawful* and *legal*; *goodwill* and *benevolence*; *help* and *aid*. The first thing, obviously, is that they have similar meanings – they're synonyms, after all, and the English language is very, very rich in synonyms. But the second thing is: although they're words of similar meaning, don't you think they *feel* different? Isn't there something deep in the character, the personality almost, of the words that triggers differing responses in you? If you went to someone in desperation, you wouldn't say, 'I need aid, please!' *Goodwill* has a more personal, practical feel than *benevolence*. *Lawful* suggests something more ethical and substantial than *legal*.

It's not just about usage: these are subtle but unmistakable shades of meaning which are the result of different linguistic roots. The first words are Anglo-Saxon or Old English in derivation, and their synonyms are Latinate or French-based. These words and thousands like them came to us from different times in our history and along different paths. Over the centuries the English language has been coloured and seasoned by those linguistic journeys. And more than that: the words tell us about our history, and where we come from.

We'll get back to the synonyms, but a bit of history first. English as we know it today was shaped principally by three periods of settlement and invasion: new peoples arrived on our shores and brought their languages with them. In the fifth and sixth centuries, it was those strangely named tribes which every schoolchild used to reel off parrot-fashion – the Angles, the Saxons and the Jutes. They sailed over from the north-western coastline of continental Europe, from present-day Denmark and northern Germany. They pushed most of the Celtic-speaking Britons north and west, mainly into

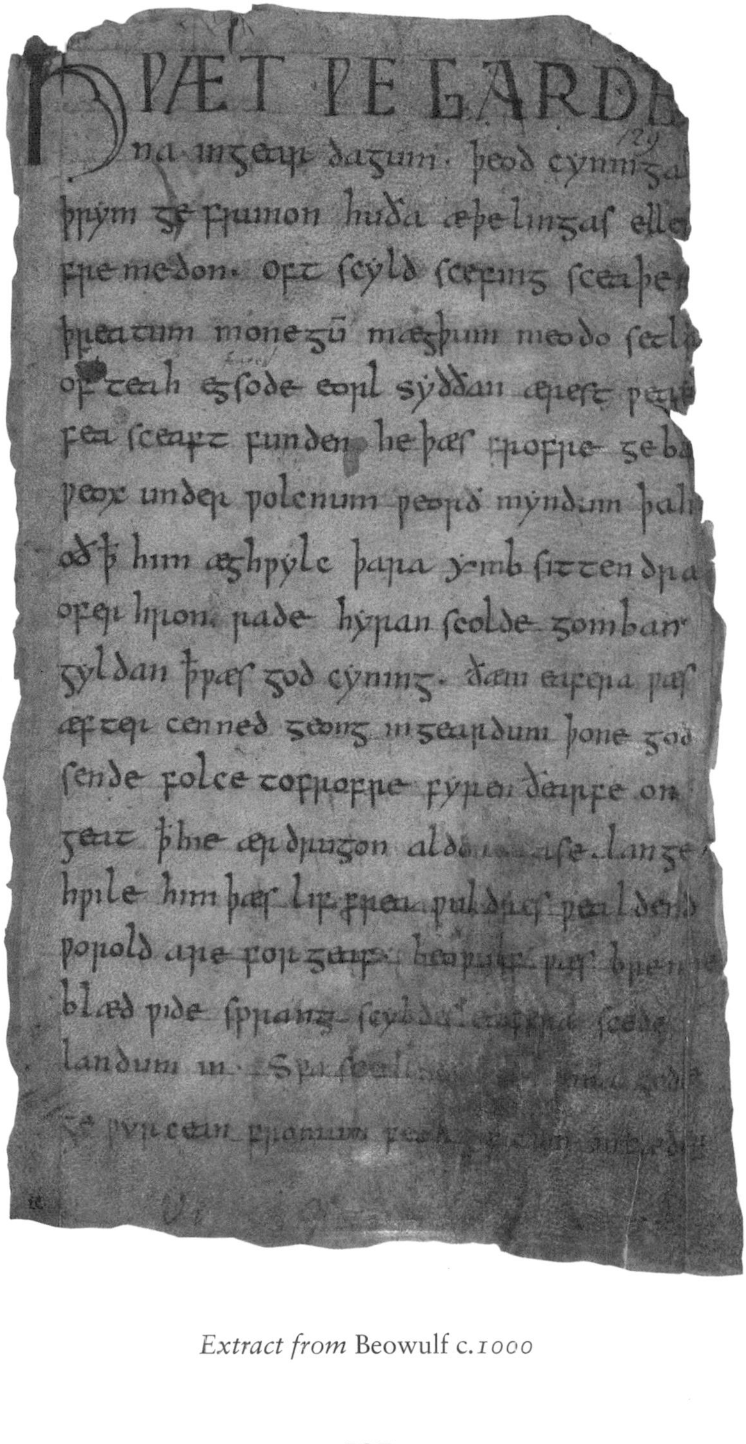
HƷÆT ƷE GARDE
na in gear dagum . þeod cyninga
þrym ge frunon hu ða æþelingas elle
fre medon . oft scyld scefing sceaþe
þreatum monegū mægþum meodo setl
of teah egsode eorl syððan ærest ƿear
fea sceaft funden he þæs frofre geba
ƿeox under ƿolcnum ƿeorð myndum þah
oð þ him æghƿylc þara ymb sittendra
ofer hron rade hyran scolde gomban
gyldan þ ƿæs god cyning . ðæm eafera ƿæs
æfter cenned geong in geardum þone god
sende folce to frofre fyren ðearfe on
geat þ hie ær drugon aldor . . ase . lange
hƿile him þæs lif frea ƿuldres ƿealdend
ƿorold are for geaf . beoƿulf ƿæs breme
blæd ƿide sprang scyldes eaferan scede
landum in . Sƿa sceal

Extract from Beowulf c.*1000*

what's now Wales, Scotland and Ireland, and their Germanic language developed into what we now call Old English. It looks and sounds just like a foreign language to a modern English-speaker (the epic poem *Beowulf* is written in Old English), but the fact is that about half of the words we most commonly use today have Old English roots, like *be*, *strong*, *water*. They are the bedrock of our language.

A few centuries later, there was another wave of invasion, and another layer added to the language mix. The Vikings were on the move. From the middle of the ninth century large numbers of Norse invaders came racing over the North Sea, raiding, trading and settling in Britain, particularly in the north and east. They came in such numbers, and were so successful, that by the eleventh century the whole of England had a Danish king, Canute. Their language was also a Germanic variety, Old Norse, and it had a deep influence on Old English. This is when very basic words like *take* and *they* come into the language, along with huge numbers of other loanwords, from the Scottish *kirk* to all these placenames with *-by* on the end, meaning 'village' or 'settlement'.

And then it was the turn of the Normans: 1066 and all that. This was the invasion which triggered enormous changes in English, as it developed into what's called Middle English (the language of Chaucer), leaving behind the inflectional system of Old English with its changing word-endings for the more or less simplified grammar we have today. Not only did English vocabulary increase hugely with the influx of French borrowings, but French became the language of the court and the ruling and business classes. For a while there was a linguistic class division, with the poor and the ordinary people speaking English and the upper classes speaking French.

There was a third language too, though: Latin. Latin was the language of the Church and of learning. Ever since St Augustine and his monks came to Britain in the sixth century, Latin (and Greek) words had been influencing English. During this trilingual period in the later medieval era, words stepped over from one language to another and left their footprints. Huge numbers of French and Latin words entered English, and English was very happy to accommodate them. The habit of borrowing and assimilating became second nature.

That's why we're swimming in all these synonyms mentioned at the start – because of the successive layers of borrowed words laid down on our language over the centuries. The majority of synonyms are from two sources – Anglo-Saxon (Old English) origin and the French or Latin origin – and they follow a distinct pattern. In general, Anglo-Saxon words are simple, essential, strong, everyday words in common usage, such as *see*, *good* and *heart*. French- and Latin-derived words are often more sophisticated, or learned or conceptual. Compare *see* and *perceive*: they mean the same, but the Latin-derived word would much more likely be used in an official, formal document.

The borrowed words are coloured by how they came into English and the status of the people who brought and used them. Latin was the language of religion and the law and learning, and French, as the language of the court, left a strong literary and cultural footprint. Our basic, concrete and close-to-the-heart words we held on to from Anglo-Saxon. Only about one-fifth of our common English words today are Anglo-Saxon, compared with three-fifths for Latin, Greek and French, but the Anglo-Saxon words are far more frequent. The hundred most frequently used English words are almost all Anglo-Saxon: *hand, foot, land, sun, cow, drink, say, have, tree, heart, life, go*. And the odd, double-word arrangement we have in English when it comes to types of meat – we call the meat by a different word from the animal it comes from – dates from the Norman period, when they brought their own words *bœuf* and *porc* for the products of the animals which had Germanic names – *cows* and *pigs*.

The synonyms, the layers, the different influences and endless loanwords all made English the flexible, subtle, vibrant language it is. They made Chaucer and Shakespeare and Joyce possible.

There have always been the purists who objected to the heavy influence of Latin, Greek and other languages on English. William Barnes was a nineteenth-century poet and philologist who argued that English needed to get back to its Anglo-Saxon roots so that ordinary people without a classical education could understand it better. So he came up with rather wonderful alternatives, like *sun-print* for *photograph*, *welkinfire* for *meteor* and the truly delicious *nipperlings* for *forceps*. He was surely outdone, though, by the American science fiction writer Paul Anderson, who set out to write a text which did not use any loanwords from other languages, particularly Latin, Greek or French. This wasn't just a simple story; his subject was atomic theory, so he had to make up a lot of new words and compounds from Anglo-Saxon roots and words, like *waterstuff* for hydrogen, *bulkbits* for molecules or the beautiful *uncleftish worldken* for atomic physics. He published *Uncleftish Beholding* in 1989, and it's as charming and inventive as William Barnes' efforts a century before.

You Say *Oïl*, I Say *Oc*

Homogenization has been with us for a long time: all empires in one way or another try to do it, and language plays a key role. Think of the Romans, the Ottomans, the British, the French – they've all imposed their language and customs on their vassal states. It's much easier to control and govern if everyone speaks the same language. That's the theory. Of course, in practice it's a different matter.

The French exported their language all over their colonies – even now we call them Francophone countries – but before they began that process they, like the British in Ireland, had to impose their language within their own borders. However, in France it was a more complicated process because there was not one version of French but two.

Occitan was once spoken by half the population of France, as well as the people of the Occitan valleys of Italy and Monaco and the Aran valley of Catalonia in Spain. It was the language of the medieval troubadours who travelled round the courts of France and Spain, singing poems of chivalry and love.

The name Occitan comes from *langue d'oc*, literally the language of *oc* – so-called because, up until the beginning of the seventeenth century, France was divided into two linguistic halves. The northern half, from around Lyon upwards, said *oïl* for 'yes', whereas the southern half said *oc*. So within France there were the *langue d'oïl* and the *langue d'oc*; the *langue d'oc* was quite different, a close cousin to Catalan, with stronger roots in Latin.

The first recorded reference to the *langue d'oc* is in an essay written at the beginning of the fourteenth century by the Italian poet Dante. In *De vulgari eloquentia* he explored the historical evolution of language. He wrote (in Latin): 'nam alii oc, alii si, alii vero dicunt oil' ('Some say *oc*, others say *si*, others say *oïl*'). The *oc* language was Occitan, the *si* language was Italian, and the *oïl* language was French.

Even as Dante was writing, the status of the *oc* language was under threat as the French royals in the north began to spread their influence into Occitan territory. In 1539, the Edict of Villers-Cotterêts decreed that the *langue d'oïl* had to be used in all legal documents and laws. The aim was to stop the use of Latin; the effect was to limit the use of regional languages like Occitan.

Outside government, the monarchy didn't seem to be that bothered about the local languages spoken by their subjects. The French Revolution changed all that. The revolutionaries argued that kings and reactionaries preferred regional languages, as they kept the peasant masses uninformed; the Republic, 'unie et indivisible' ('one and indivisible'), couldn't be achieved if the spread of new ideas was hindered by patois – a derogatory term used for the countrified, backward languages of non-Parisian French-speakers.

On 27 January 1794, Bertrand Barère, a member of the National

'Speak French Be Clean' daubed on the wall of a school

Convention who had presided over the trial of Louis XVI, argued in favour of one national language:

The monarchy had reasons to resemble the Tower of Babel; in democracy, leaving the citizens to ignore the national language, unable to control the power, is betraying the motherland . . . For a free people, the tongue must be one and the same for everyone . . . How much money have we not spent already for the translation of the laws of the first two national assemblies in the various dialects of France! As if it were our duty to maintain those barbaric jargons and those coarse lingos that can only serve fanatics and counter-revolutionaries now!

A few months later, Abbé Grégoire, priest and revolutionary leader, presented his infamous 'Report on the necessity and means to annihilate the patois and to universalize the use of the French language'. Grégoire denounced the fact that fewer than 3 million of France's 25 million citizens spoke *la langue nationale*; French, he argued, must be imposed on the people of France, and all other dialects must be eradicated.

The revolutionaries ran out of time and money before they could 'annihilate' local languages, but the modernizing of France during the nineteenth century did the job just as well. Boundary changes and the creation of new *départements* cunningly undermined regional identity; the introduction of the railways opened up isolated areas to travel and commerce and the French language; compulsory military service meant soldiers from around the country had to speak French to be understood; but, crucially, the introduction of compulsory education in the latter half of the nineteenth century sounded the death knell for Occitan and other minority languages. Hand in hand with the progressive ideas of education for all children came the tools to suppress local languages. The Occitans have a word for the policies which successive French governments used to eradicate minority languages: *la vergonha* – 'the shame'.

Primary children caught speaking Occitan or Breton or Catalan or any non-French language in class or in the playground were made to wear or carry *la symbole* – or *la vache* ('the cow'), as pupils referred to it – during the school day. The *symbole* might be a wooden clog or a horseshoe or a slate with a description of the felony written on it. Offenders faced extra homework or sometimes corporal punishment at the end of the day.

The policy was highly effective. A Breton remembers the so-called clogging:

> *My grandparents speak Breton too, though not with me. As children, they used to have their fingers smacked if they happened to say a word in Breton. Back then, the French of the Republic, one and indivisible, was to be heard in all schools, and those who dared challenge this policy were humiliated by having to wear a clog around their necks or kneel down on a ruler under a sign that read: 'It is forbidden to spit on the ground and speak Breton'. That's the reason why some older folks won't transmit the language to their children: it brings trouble upon yourself. (Nicolas de la Casinière,* Ecoles Diwan, la bosse du breton*)*

France wasn't the only country to employ *symboles* in their schools. In Wales in the late nineteenth and early twentieth centuries, schoolchildren were forced to wear a Welsh Not – a piece of wood inscribed with the letters WN – around their neck if they spoke

A Welsh Not, hung around the neck of the first child caught speaking Welsh

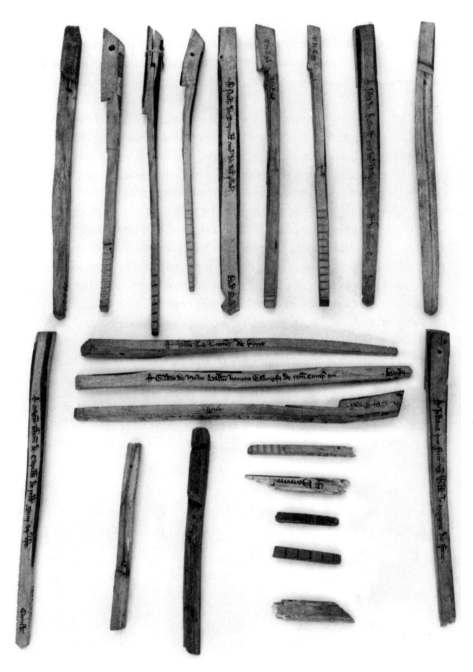

*Irish tally sticks, worn around the neck and marked with a notch
whenever a child spoke in Irish*

Welsh and not English. Each morning the wooden board was hung around the neck of the first child caught speaking Welsh, who handed it to the next child who offended. The poor child wearing the board at the end of the school day was punished. In Ireland, the tally stick or *bata scóir* was introduced into the classroom. Each child had to wear the stick on a piece of string around their neck, and every time Irish was spoken, the teacher would add a notch. A corresponding number of whacks was dealt out to the child at the end of the day.

It took until the 1950s in France for the policy of language repression in primary schools to end and for the French state to recognize the right of regional languages to exist. Yet unlike Spain or Wales, for example, where Basque and Welsh have been given equal status as a national language, there are no signs that French regional languages will ever be given equal footing. In 1992, the French constitution was revised so that 'the language of the Republic is French'. France remains one of the few European states which refuses to ratify the European charter for minority languages, giving legal status to its regional language groups.

In 1851, at the time of the French education reforms, about 40 per cent of the country spoke the *langue d'oc*. Today, much less than 1 per cent are fluent speakers. Occitan can only be taught in schools as a foreign language. Its future looks bleak.

Mistral Revives Provençal

As French governments sought to wipe out Occitan and other regional languages, a young poet in the south of France, Frédéric Mistral, determined to revive Provençal, a dialect of Occitan spoken in Provence. 'Instead of setting their ambitions on the baubles of Paris or Madrid, we want our daughters to continue to speak the language of their cradle.'

In 1854, Mistral and six friends started the Félibrige, an association dedicated to the revival of the Provençal language and customs. Mistral spent twenty years working on a dictionary of Provençal, *Lou Tresor dou Félibrige*, but his greatest contribution to the Provençal revival was his poetry; he published

Frédéric Mistral was dedicated to the revival of Provençal

his first poem, *Mireio*, the story of a rich farmer's daughter and her love for the son of a poor basketmaker, in 1859. Three other long narrative poems followed, as well as volumes of short stories, lyrics and memoirs. In 1879, Mistral's *A Na Clemenco Isuaro* was first recited in Toulouse at the Jeux Floraux, the world's oldest literary competition. Centuries before, during the time of the troubadours, it had celebrated Occitan poetry, but after the Edict of Villers-Cotterêts in 1539, the only contributions were in French. Now, under Mistral's urgings, Occitan was readmitted on a par with French. The competition is still held in Toulouse on 3 May each year.

In 1904, Mistral was awarded the Nobel Prize for Literature, 'in recognition of the fresh originality and true inspiration of his poetic production, which faithfully reflects the natural scenery and native spirit of his people, and, in addition, his significant work as a Provençal philologist'. Mistral and the Yiddish author Isaac Bashevis Singer are the only two writers to have won the award in a minority language.

Mistral was fêted as a great Provençal – and French – poet. There's a delightful story of William 'Buffalo Bill' Cody arriving at Mistral's farm near Arles one day in 1905. Mistral had visited Buffalo Bill's Wild West Show during one of its tours of the south of France. The showman apparently returned the visit and left the Nobel laureate a present – a dog which Mistral called 'Pan Pardu', Provençal for French toast. Stories about the two may have been fuelled by the uncanny similarity between the two goatee-sporting artistes.

Mistral continued to lead the Félibrige movement until his death in 1914.

Parlez-vous Franglais?

The crusade for the supremacy of the *langue d'oïl* – 'Paris French' – over the regional patois has clearly been won. But in an ironic twist, which would no doubt have Mistral singing his Provençal songs in his grave, the official French language now finds itself in a battle against the rest of the world. French, the language of culture and diplomacy for over 200 years, is under threat as never before.

There was a telling moment a few years ago when the then French President, Jacques Chirac, stormed out of an EU summit meeting after a French business leader addressed delegates in English. President Chirac told reporters, 'I was profoundly shocked to see a Frenchman express himself in English at the table.' For his part, the businessman said he chose English 'because that is the accepted business language of Europe today'. President Chirac's walkout (and his insistence, despite being fluent in English, of speaking French throughout a one-to-one dinner with the non-French-speaking President Bush) provides a vivid illustration of how sensitive the French – or at least the French establishment – are about the decline of the language. And it's not a recent concern.

On the banks of the Seine in Paris the imposing Institut de France is the home of the Académie Française. It was founded in 1635 by Cardinal Richelieu with the express purpose of keeping the French language pure. Just as Paris French, the *langue d'oïl*, was being imposed as a language of state within France, so the intention was that it should prevail in the diplomatic field to rival Latin as the language of Europe. Since then, this Star Chamber for words has survived everything from the Revolution to Nazi occupation. A linguistic jury is made up of forty members known as *immortels* (immortals) who hold office for life and are charged with establishing how French is written and spoken. The recommendations of the jury have no legal power but are taken very seriously. And the French they advocate has no truck with regional dialects such as Basque or Occitan. As Marc Fumaroli, one of the most outspoken of the *immortels*, explains: 'But you know, what they, the regional languages, have lost is not too much, and in compensation they have been able

The Académie Française regulates correct usage of the French language

to participate in one of the most wonderful conversations possible, the conversations in Paris.' No regrets there, then. Members are tasked with working on a dictionary of the French language, known as the *Dictionnaire de l'Académie française*. Eight editions have been published since 1635; the first one took fifty-nine years to complete, and Academicians have been working on the ninth since 1935.

Since the end of the Second World War and the growing domination of the English language in business, diplomacy and culture, the *immortels* have increasingly focused on trying to prevent an influx of *franglais* – English terms used in French – by choosing or inventing French equivalents.

Some recommendations have failed miserably: *le weekend* has never been threatened by the proposed *la vacancelle*; *le shopping*, *un parking* and *le discount* are all firmly part of French everyday speech. However, the attempt to banish the use of the word 'email' by proposing *un courriel* (from *courrier électronique*) has met with partial success. *Courriel* is used in written French, but internet users talk about *mail* or *mel*. Other loanwords like 'walkman' and 'software' have been replaced with *baladeur* and *logiciel*.

The Academy has sometimes been criticized for being too conservative. The increase in the number of working women, for instance, has caused problems, as many of the French nouns for the professions don't have feminine equivalents. An attempt by Lionel Jospin's government in 1997 to refer to a female minister as *Madame la ministre* was torpedoed by the Academy, which insists on the traditional use of the masculine noun, *le ministre*, for either gender. The Toubon Law (nicknamed the 'Loi Allgood' – 'tout bon') mandates the use of the French language in official government publications, in all adverts and in all workplaces. A percentage of the songs on radio and television must be in French as well. The annual satirical award of the Prix de la Carpette (literally the Rug or Doormat Prize) goes to the person or institution that has given the best display of 'fawning servility' by insinuating the 'accursed English language' into France. It's not sponsored by the Academy, although one of its members is on the jury. Nominees have included France Telecom for publishing titles such as *Business Talk* and *Live-Zoom* and the president of the French Football Federation for adopting the Jackson 5 hit 'Can You Feel It' as the anthem of the French national team.

There are some truly awful examples of mangled *franglais* which make you wish the Academy had a bit more clout. People who like to stay at home are said to *faire du cocooning*; a failure is *un looser*; a makeover is *un relooking*; and fine cooking is *le fooding*; and if you dress sexily, you've gone for *l'image total destroy*.

France isn't ready to throw in the towel on the linguistic world stage yet. A British spectator at the 2012 Olympic Games in London may be surprised to see that all the official notices displayed around the capital will be in French as well as English. And that the announcement as the medal winners climb on to the podium will be in French first, then English. French is the first tongue of the International Olympic Committee – the founder of the modern Olympic Games was French after all, and the IOC headquarters are in French-speaking Geneva.

It's easy for purveyors of the English language to mock the French attempts to hold on to their own. Let's face it, spoken English dominates the world because it's the language of the USA. If history had been different, if the French colonials had dominated across the Atlantic, then the British would undoubtedly be fighting a similar linguistic battle themselves.

The Survival of Basque

In contrast to the demise of so many regional languages in France, the Basques in Spain have succeeded in preserving both their language and their culture. Decades of armed struggle by the seperatist organization ETA may have alienated large sections of the Basque population, but it has also won them rights that are the envy of the Basques over the border in France. It is possible now to go from nursery school to university in an entirely Basque medium.

San Sebastián is one of those magical Atlantic coast cities battered by the waves and offshore winds that bring a salty taste to everything. It is a handsome city, purveyor of a fine International Film Festival and home to some ridiculously good cooking. It is the cultural centre of the Basques.

The three-Michelin-starred chef Juan Marie Arzak runs a restaurant in the city with his daughter Elena. Juan and Elena talk passionately about their Basque identity and how they celebrate it through their cooking. Whilst the DNA is Basque, it is married to an intensely modern, cosmopolitan and experimental style which is happy to infuse Persian limes with Vietnamese coffee beans to enhance a traditional Basque sauce. Not only is the food exquisite but it exemplifies the confidence they have in their own identity. They will not be subsumed, as history will testify. And their language, even more than their cuisine, is at the heart of it all.

The language of the Basque people is related to no other language on earth. Although the Basque country straddles the border between Spain and France and is surrounded by Indo-European languages, Basque is what we call a language isolate. It was spoken in Europe long before Indo-European people began migrating across the continent and is very ancient – maybe tens of thousands of years old. Linguists have puzzled for years over the origins of Basque, suggesting that it developed with the Berbers in northern Africa or in the Caucasus or on the Iberian peninsula. Some linguists suggest it may even have evolved as far back as the Stone Age, pointing out that the Basque word for axe comes from the root word *haitz*, which means stone or rock. Other more wacky suggestions have made the Basques the lost thirteenth tribe of Israel or survivors of Atlantis. One theory even has Adam and Eve speaking Basque.

The Basque language wasn't written down until the sixteenth century. It has a fiendishly complex grammar and it looks alien, certainly to the European eye, full of *z*s and *x*s. Just counting from one to ten is a challenge: *bat bi hiru lau bost sei zazpi zortzi bederatzi hamar*. Its obscurity proved a rather useful secret weapon during the war in the Pacific in the Second World War. The US army bamboozled the Japanese intelligence services by sending coded messages in Basque (as well as Navajo, Iroquois and Comanche). Basque-speaking marines were used to send and receive the transmissions; the phrase to signal the beginning of the Battle of Guadalcanal in 1942 was 'Egon arretaz x egunari' ('Heed the x day').

The language of Euskara, as the Basque people call it, is spoken today by around 600,000 people in seven traditional regions – four in Spain and three in France. While the Basque language in France has been subjected to the same fate as all the French regional languages and has no official status, the Spanish Basques have enjoyed a much greater degree of political and cultural autonomy. Euskara did suffer after the Spanish Civil War, when the language was outlawed under the dictatorship of General Franco. Chef Juan Arzak remembers being punished for speaking Euskara in school. A Basque nationalist movement grew under Franco; since his death, Spanish Basques have regained some self-rule powers, but the nationalist movement continues – in its most extreme form as the terrorist organization ETA.

Today, Euskara has co-official status with Spanish in the region, and the number of speakers is growing. Children can choose to attend schools where all lessons are in Euskara and Spanish is taught as a second language. There's a Euskara-language television and radio station, newspapers and books in Euskara. A Royal Academy of the Basque language is charged with watching over Euskara, protecting it and establishing standards of use. It's a daunting task. The younger generation of Basques tends to speak *euskañol*, an informal mixture of Euskara and Spanish. But people like Chef Arzak aren't worried. The Basque food he cooks is a mixture of the traditional and the avant-garde and draws on dishes and ingredients from other food cultures; in the same way, the Basque language is able to incorporate words and expressions from other languages. It's a pungent metaphor for a unique language.

The Story of Hebrew

The Basques are fortunate in having mountains, rivers and the sea to define their borders and repel invaders. This has without doubt helped them to preserve their language when many other minority languages in Europe have withered. The Diaspora of the Jews could easily have resulted in the disappearance of their language, as they were absorbed into other cultures, but their resistance against anti-Semitism and the strength of their religious life meant that Hebrew

אִישׁ אֶחָד מִן הָרָמָתַ֫יִם

הָרָמָתַיִם צוֹפִים מֵהַר

אֶפְרַיִם וּשְׁמוֹ אֶלְקָנָה

בֶּן יְרוֹחָם בֶּן אֱלִיהוּא

בֶּן תֹּחוּ בֶן צוּף אֶפְרָתִי

וְלוֹ שְׁתֵּי נָשִׁים שֵׁם

אַחַת חַנָּה וְשֵׁם הַשֵּׁנִית

פְּנִנָּה וַיְהִי לִפְנִנָּה

יְלָדִים וּלְחַנָּה אֵין יְלָ

 דִים וְעָלָה הָאִישׁ

Samuel extracted from a fourteenth-century Hebrew Bible

was kept alive during centuries of exile until, in the aftermath of the Holocaust, the state of Israel was created, and Hebrew was adopted as the national language.

Whatever one thinks of Israel, its achievment in inculcating a sense of identity is without doubt one of the success stories of the postcolonial era. The country had a tricky birth, a difficult childhood and a tempestuous adolescence. It has been an extraordinary journey to nationhood, and language has been at the core of keeping it together.

Dr Ghil'ad Zuckermann is a leading Israeli linguistician, as people who study linguistics are properly called to differentiate them from linguists. He tells a story of five Jews who influenced the world. It goes like this:

Moses said, 'The Law is everything.'
Jesus said, 'Love is everything.'
Marx said, 'Money is everything.'
Freud said, 'Sex is everything.'
Einstein said, 'Everything is relative.'

He sees the punchline as a lesson on language, specifically modern Hebrew, which he calls Israeli. 'Israeli is a very complex language,' he says. Its revival has been a 'relative success', a mish-mash of old Hebrew and different bits of other languages, a hybrid affair – just like everything else in life.

Not everyone's laughing, though. Ghil'ad, who was born in Tel Aviv and is the Professor of Linguistics and Endangered Languages at Adelaide University in Australia, is a controversial figure in Israel. His line on Hebrew annoys a great many people: he argues that the Hebrew of the modern state of Israel is not, as most linguists believe, a revival of the ancient language which hadn't been spoken for nearly 2,000 years. Israelis are not speaking like the prophet Isaiah, he insists. Neither does he accept the theory of relexification – that modern Hebrew is, in essence, Yiddish (the language most of the early settlers spoke) with Hebrew words. Not true either, he says. Ghil'ad argues that Hebrew is a synthesis, a language with more than one parent, a beautiful mongrel, a hybrid. It is genetically both Indo-European and Semitic. Ancient literary Hebrew and Yiddish are the primary contributors, and there is an array of secondary influences, like Arabic, Russian, Polish, German and Ladino (Judaeo-Spanish), all languages spoken by the Hebrew revivalists.

Ghil'ad's thesis challenges the traditional linguistic wisdom that, with a bit of tinkering and adding of new words, modern Hebrew sprang directly out of the ashes of ancient liturgical and literary Hebrew, which had once been spoken all over the Jewish world. It also annoys politicians and scholars, who accuse him of pushing a political, post-Zionist agenda. The early Zionists emphatically linked the revival of Hebrew to their claim of an ancient birthright in the land of Israel, so the modern state doesn't appreciate a linguistician who says that Israelis aren't actually speaking the language of the prophets.

These arguments matter because the story of Hebrew is not just the story of a language. It's about history and nationhood and identity – and the battle for survival.

For centuries Hebrew was the language in what is now Israel-Palestine, generations before a man called Jesus ever walked the streets of Jerusalem. But after the Diaspora, when the Jews scattered throughout Europe following the revolts against the Romans in the first and second centuries AD, Hebrew died out as a spoken language. It became a linguistic Sleeping Beauty, read, recited and remembered only in the Torah, the rabbinical tradition and Friday-night suppers in Jewish homes. In the nineteenth century, the movement for a Jewish homeland, a return to Zion after 2,000 years of persecution, began to take hold. And with it came the dream of reviving the language of the ancients.

Rishon LeZion is a bustling suburb of Tel Aviv. It was founded in 1882, one of the first Zionist settlements, and in 1886 the first Hebrew school was created there. This was the brainchild of the prime mover behind the revival of Hebrew, Eliezer Ben Yehuda, a Jewish lexicographer and newspaper editor. He had learned written Hebrew as part of his religious upbringing, and at one time thought of becoming a rabbi, but instead he became increasingly interested in the nationalist movements in Europe and ideas of self-determination. He believed that, if Jews returned to their ancient land and began to speak their own language, they could once again become a nation. For him, Hebrew and Zionism were one: 'The Hebrew language can live only if we revive the nation and return it to the fatherland,' he wrote. So, in 1881, he moved with his family to Palestine, then part of the Ottoman Empire.

He was the original Man with a Mission, writing newspaper articles about reviving Hebrew and speaking only Hebrew with every Jew he met, despite the lack of current vocabulary. The Hebrew he used was the language of the synagogue, of psalms and prayers and the Torah, and a flowery literary tradition which was heavily mixed with Aramaic. But Eliezer was passionate in his conviction that, with training and practice and the necessary creation of new words, Hebrew could and should be resurrected as a spoken language. He was a bit of a tyrant and insisted his family speak only Hebrew, not Yiddish or their native Russian. His son, Ben-Zion, was not allowed to come into contact with any other language, and Eliezer boasted that, as a result, his son was the first native speaker of Hebrew. Ben-Zion wrote later in his autobiography that he was sent to bed if foreign visitors came to the house, and that when Eliezer came home one day and found his wife absent-mindedly singing a Russian lullaby to Ben-Zion, there was a furious shouting match.

Members of the Hebrew Language Council in 1912

Eventually, it was the successive waves of Jewish immigrants from all over the world which secured the revival of spoken Hebrew. In the last decades of the nineteenth century and early ones of the twentieth, thousands of Jews flooded into Palestine, mainly from Eastern Europe and Russia, and Hebrew schools were set up, where the medium of teaching for everything was Hebrew.

Of course, the teachers weren't native Hebrew-speakers either. They were mostly Ashkenazi Jews, and their language was Yiddish. They tried to teach the best Hebrew that was available to them, but it was very difficult. 'In a heavy atmosphere, without books, expressions, words, verbs and hundreds of nouns, we had to begin teaching,' wrote one teacher from the 1920s. 'We were half-mute, stuttering, we spoke with our hands and eyes.' When it came to making new words, 'everyone, of course, used his own creations'.

It's a key point in Ghil'ad Zuckermann's argument for the hybridization of Hebrew. The early settlers and teachers – what he calls 'the founder population' – spoke Yiddish or Polish or Russian and, he says, they could not rid themselves of their native grammatical structures. And a Yiddish-speaker pronounced Hebrew differently from, say, a Sephardic Jew from Yemen. Ghil'ad argues that this early influence of Yiddish-speakers on Hebrew left an indelible trace on its structure and word order. 'Yiddish speaks itself beneath Israeli,' he says. 'Had the language revivalists been Arabic-speaking Jews [from Morocco, for example], Israeli Hebrew would have been a totally different language – both genetically and typologically, much more Semitic.'

By the end of the First World War, Hebrew was consolidating its position as the national language-to-be. There was a movement in favour of Yiddish, the lingua franca of the middle European Jew, but it could not throw off the taint of the *shtetl*, the pogroms and the poverty. The Zionists wanted to re-establish the link to the older story, the ancient Jewish history and the centuries of nationhood, and Yiddish was despised as second-class and polluted.

In 1909, the first Hebrew city, Tel Aviv, was established. The entire administration of the city was carried out in Hebrew, and people were forced to speak Hebrew. Street signs and public announcements were written in Hebrew. Hebrew was so dominant in Tel Aviv that, in 1913, one writer announced that 'Yiddish is more *treif* [non-kosher] than pork. To speak it a person needs great courage.' Yiddish-speakers were shouted at in the street – 'Jew, speak Hebrew' – or even beaten up by gangs of men from the creepy-sounding Defenders of the Language Brigade.

In 1921, Hebrew became one of three official languages in British-ruled Palestine (along with English and Arabic), and then in 1948, with the founding of the state of Israel, it was made an official language along with Arabic. Today it is the most widely spoken language in Israel.

But what if Israel had chosen Arabic as its main national language? Would it have been a means of reconciliation between Jew and Arab? Ghil'ad Zuckermann wonders, if Israelis and Palestinians had literally been able to understand each other, whether it could have changed the political story and prevented the cumulative violence and antagonism of the years after 1948. The two languages are so close, Hebrew and Arabic (although Ghil'ad says Arabic has far better curses and swear words), but not close enough to overcome what are essentially ancient tribal differences and conflicts which go all the way back to the story of the separation of Isaac and Ishmael. To this day the word in Hebrew for Arab is *Ishmael*.

The reason for Hebrew's success is the other side of the same coin: it had a fervent, strong ideology of a national movement. People wanted a common language to unify them, all these Jews coming from all over the world and speaking different languages. Hebrew succeeded through political will, and a strong identity which both secular and religious Jews have used for their own purposes.

Yiddish: Not Just for Schmucks

Yiddish has been called, variously, the language of the ghetto and the *shtetl*, a barrier to assimilation, a mutilated and unintelligible language without rules, the mother tongue, the true language of the Jewish people – and a dozen other things, some complimentary, some not. At its height before the Second World War, it was understood and spoken by an estimated 11 million Jews, but it lost the battle with Hebrew to become the national language of Israel, and the number of Yiddish-speakers has dwindled rapidly since. And yet it survives, albeit spoken principally in the United States and by the ultra-orthodox in Israel. As the writer Isaac Bashevis Singer said, 'Yiddish has not yet said its last word.'

In fact, Yiddish has borrowed and assimilated so many words from other languages, and given so many in return, especially to English (think schmalz, chutzpah, schmutz, bagel, shmooze), it's easy to forget it's not a sort of Jewish-American form of English. It's a proper language, a hybrid of medieval German and Hebrew, using the Hebrew alphabet, with dashes of Aramaic and Slavic languages thrown in. It developed among the Ashkenazi Jews living along the Rhineland around the tenth century, and then spread through central and eastern Europe. The ancient tongue of the Jews in the kingdoms of Israel and Judah, Hebrew, had been in decline as a spoken language since the Jews were defeated by the Babylonians in the sixth century BC, and well before the time of Jesus it had been replaced by Aramaic as the Jewish vernacular. But Hebrew remained the written language of religion, of the liturgy, the Torah and the scholars.

Yiddish – which means 'Jewish' – takes about three-quarters of its vocabulary from medieval German, but it has borrowed liberally from Hebrew and from many other lands and cultures where the middle European Jews lived. It has tremendous richness of expression, and the vigour and subtlety of a mongrel creation, like English, which amasses words from all over the place.

It developed a rich vocabulary to express the human condition, often using humour to rail against suffering and the absurdities of life, which was a useful antidote to the pogroms, persecution and expulsions. Many of the terms have found their way into English, because there is no equivalent English word which can convey the depth and precision of the Yiddish. There's a tremendous humanity and irony in the sheer range of words to describe the human character in all its failings and strengths. For example, what other language would distinguish between a *schlemiel* (someone who suffers through his own actions), a *schlimazl* (a person who suffers through no fault of his own) and a *nebach* (a person who suffers because he makes other people's problems his own)? There's an old joke which explains the distinction: the *schlemiel* spills his soup, it falls on the *schlimazl*, and the *nebach* cleans it up.

Yiddish has been a potent force in Jewish identity and the struggle for survival. But as Jews became assimilated into the local culture, particularly in Germany from the late 1700s and into the 1900s, Yiddish began to decline. It was criticized by Jews, particularly those most influenced by the ideals of the Enlightenment, as a barbarous ghetto jargon, a barrier to Jewish acceptance in German society.

In 1908, the first Yiddish Language Conference was held in Czernowitz, Bukovina (now Chernivtsi, Ukraine). It was supposed to be about standardizing the language and promoting and celebrating its thriving literature and drama, but it became a fierce argument

Ticket to the first Yiddish Language Conference, 1908

about the status of Yiddish relative to Hebrew, and which could truly lay claim to being *the* national Jewish language. Politics, history, tradition and the future all got mixed up with debates about language, and there was a great deal of passionate speech-making, including a keynote address from the writer I. L. Peretz. 'There is one people – Jews,' he declared, 'and its language is – Yiddish. And in this language we want to amass our cultural treasures, create our culture, rouse our spirits and souls and unite culturally with all countries and all times.'

It's interesting that the creator of Esperanto, Ludovic Zamenhof, was also part of the debate about Yiddish and Hebrew, but he took a different and independent view. He had created Esperanto to be a uniting force for good in a disparate world. He believed that a new, international language could bring people of different cultures and nations together; everyone would have their own languages, but Esperanto would be the vehicle for equality and understanding and peace among everyone. Later in life, he took his vision even further; he dismissed Yiddish and Hebrew, and a physical Jewish state in Palestine, and argued that Esperanto was the language to promote a broader spirituality, and bring together Jews the world over in a new kind of mystical humanism.

As the Zionist movement gathered momentum, more and more Jews from around the world, speaking many different languages, began to return to Palestine, and, in the end, Yiddish lost out to a revived and modernized Hebrew. After the Second World War it was even more tainted with the memories of the Holocaust, and the modern state of Israel turned its back on it. The once thriving language of middle Europe survives mainly amongst the elderly Jewish communities in New York, Florida and California and, of course, in the world of comedy and entertainment.

The Friar's Club in midtown Manhattan looks like a medieval castle inside, but it's a castle owned and controlled by the court jesters. Founded as a private members' club in 1904, it has been a bastion for entertainers, and especially comics, for over a century. Will Rogers, Jack Benny, Jerry Seinfeld, Billy Crystal – well pretty much every major and minor American entertainer – have graced the oak-panelled rooms. And, of course, this being New York, it is the Jewish comedians who have dominated the scene. Stewie Stone is a veteran of the so-called 'Borscht Belt'. Also known as the Jewish Alps, the Borscht Belt was like a mini Las Vegas that flourished between the 1930s and the 1960s in the Catskill Mountains in upstate New York. It was a place where the predominantly Jewish population of NYC would go for their summer vacations and listen to the quickfire repartee of stand-up comedians, many of whom would become household names – Mel Brooks, Woody Allen, George Burns, Jackie Mason, Phil Silvers and Joan Rivers, to name just a few, and, of course, the irrepressible Stewie Stone, whose Yiddish humour is rooted in the fertile comic soil of Flatbush in Brooklyn, where he grew up.

Jewish comedians Woody Allen and Billy Crystal with Elizabeth Shue in Deconstructing Harry, *1998*

Stewie, with his broad Brooklyn accent, is a powerful advocate for Yiddish. Holding court around a pool table at the Friar's Club, he expounds: 'It's a great language for saying the man's a dick. As the snow is to Eskimos, so schmuck is to the Jew. The definition of a schmuck is a guy who gets out of the shower to take a piss. A *schlemiel* is a loser, and a *nebach* is like a waste of time. You can talk to a *schlemiel* – I'll sit and have a drink with him – but a *nebach* – you don't want nothing to do with a *nebach*.'

Ari Teman, a young Brooklyn comic, concurs: 'The *nebach* is the guy getting the water when you're going on a football team,' and, quick as a flash, Stewie responds: 'For an older Jewish audience, if you said "ass" on stage, you were dirty. "Kiss my ass", you are dirty. But if you said "kish mein tuchas" there, that was clean, it was accepted. You could curse in Yiddish and not offend anybody.'

Ari adds: 'I sent a rabbi a joke, and he wrote back the acronym "LMTO". Normally it's "laughing my ass off", but he replaced it with a "tuchas". "Laughing my tuchas off", it's a Yiddish acronym. We're instant messaging Yiddish acronyms now.'

Stewie continues: 'When one Jew meets another Jew he's a *landsman* [literally countryman], meaning we're of the tribe. No matter where you are. So two old Jewish men were talking, and one said, "Morris, *bei* you [according to you], what is a good *Shabbat*?" And the other said, "A good *Shabbat*, *bei* me, is I get up, I go to school, I *doven* [pray], I come back, I sit with my grandchildren, and we eat, and I go back to school, and we *doven*. *Bei* me, that's a good *Shabbat*. Seymour, *bei* you, what's a good *Shabbat*?" The other man says, "I get up in the morning, I get on a plane, I fly to Vegas, I get off the plane, I go to the crap table, I shoot crap, I win 10,000 dollars, I go up to the room, there's a chick waiting for me. *Bei* me, that's a good *Shabbat*." And the other guy says, "Good *Shabbat*? *Bei* me, that's a great *Shabbat*!"'

And of course it's so important that that joke is in Yiddish, that you say *Shabbat* and you don't just say Sabbath.

There's a joke about the *shtetl* in Poland or southern Russia somewhere, a little village, where the Jews live in perpetual fear of a pogrom. There's a story in the newspaper of a young teenage girl being murdered, and they're terrified it's going to be a Christian girl and that the local community will blame the Jews. Suddenly one of their number runs in and says, 'I've got fantastic news, the murdered girl was a Jew.' As a joke it's both terrible and funny, because the Jews were persecuted beyond language. It's almost unspeakable what happened. And yet always there's this realist strain of humour.

Stewie develops this serious point: 'Yiddish is basically our soul, and the kids are starting to go back to it. It's our heart, our soul. Jews are emotional people. We love to cry, and in Yiddish you can cry. You can't cry in Hebrew too easily. Judaism is not a religion, it's a way of living your life. And Yiddish is a way of feeling your life because every time you use these words . . . it's taking us back to where we came from, that we all share together. We're all the same Jew. We're not a rich Jew or a poor Jew.'

Stewie doesn't believe that Hebrew can fulfil this function.

'I grew up, and I was *bah mitzvahed*, but we didn't talk Hebrew. We never thought of talking Hebrew, it was the language of the Temple, it was something we had to learn. Whereas Yiddish, I would hear my grandparents and my parents talk Yiddish. But they didn't want me to hear. *Zug gornisht!* – it means "Don't say anything" – the kid is listening. And we would try and translate what they were saying.'

So Yiddish was the language of emotion and sex and failure and hilarity, whereas Hebrew was the language of seriousness and ceremony and solemnity.

Stewie sums it up: 'Hebrew comes from the vocal chords, and Yiddish comes from the heart.'

Esperanto: Language and Utopia

All languages are, of course, invented. They are the product of the human mind and our boundless creativity and desire to communicate. But usually they evolve more organically than modern Hebrew, and often with remarkable alacrity, which is especially true with patois, creoles and slang. But there is a whole raft of entirely invented languages. Many have an idealistic basis, and from the seventeenth century on, when Latin began to lose its place as the lingua franca of intellectual writing, there was a stream of attempts at creating more rational and logical languages that would not only be universal but also better able to describe the burgeoning advancements in science.

John Wilkins, inventor, scientist and founder of the Royal Society, spent ten years trying to create the perfect language, free of ambiguity and so precise it could be used to describe these truths. In 1668, he presented his work, *An Essay Towards a Real Character and a Philosophical Language*, to the Royal Society. Hooke, Newton, Locke and Leibniz all read it with interest, and there was heated debate in the coffee houses of London as to the degree to which language got in the way of understanding objects and whether Wilkins' method, where the word tells you exactly what it signifies rather than being an arbitrary sign, was really an improvement. What his language did prove was that it was possible to create a perfectly sound, logical language from scratch. The entire universe was divided into categories, and everything belonged to one of these by ever more detailed description. So 'shit' becomes *cepuhws* – 'a serous and watery purgative motion from the gross parts of the guts downward' – a word that is created from four different categories expressing 'motion', 'purgation', 'gross' and 'of vomiting'. You can see why it never caught on. Languages by their very nature are messy and fluid. His analytical methodology, however, did lay the foundations for the thesaurus and later for taxonomy, the science of classifying things.

Since Wilkins' magnificent failure there have been countless attempts to invent the perfect language. The nineteenth century offers a cornucopia of new languages, but most of them, rather than re-inventing the wheel, sensibly used existing languages as bases. Pirro's Universalglot, Schleyer's Volapuk and then the most famous and successful of them all, Esperanto, all took elements of European languages as their source.

Esperanto was the creation and life's passion of a nineteenth-century Russian-Jewish ophthalmologist called Ludovic Lazarus Zamenhof. He was born in 1859 in the city of Bialystock, which is now in Poland but was then part of the Russian Empire. It was a multilingual, multicultural city made up predominantly of Jews, with Polish, Russian, German and Belarussian minorities. Zamenhof grew up speaking Russian, Yiddish, Polish and German, deeply secure in the culture and history of his Jewishness, while living cheek-by-jowl with the competing languages and nationalisms of his neighbours.

From an early age he was fascinated by the possibilities of a completely new language which would break down national and cultural barriers and foster understanding and peace. It would be a new lingua franca with none of the colonial baggage of English or French or Russian because it would be neutral and borderless, and it would be simple enough to learn in far less time than any other foreign language.

Ludovic Zamenhof,
inventor of Esperanto

Throughout his time at school and university, Ludovic pursued his dream of creating this new, unifying language. 'I was taught,' he wrote, 'that all men were brothers, and, meanwhile, in the street, in the square, everything at every step made me feel that men did not exist, only Russians, Poles, Germans, Jews and so on.' His father was not impressed. He spoke to the headmaster at school about Ludovic's ideas and was told that that his son was lost for ever, and that his work was the surest symptom of the onset of an incurable madness.

Zamenhof studied medicine but lasted only a few months as a GP. He didn't like it and decided to specialize in ophthalmology instead. By the time he set up business in Warsaw in 1887 he had married Clara Zilbernik, the daughter of a rich soap manufacturer, who brought a dowry of 10,000 rubles with her. This was supposed to support the family, but instead Ludovic asked his father-in-law if he could invest half of it in the dream which still drove him: the promotion of his new international language. With the money he published his first booklet in 1887, a basic grammar of Esperanto, in Russian. He used the pseudonym Doktoro Esperanto (Doctor Hopeful), and the book was so successful that Polish, French and German versions came out before the end of the year. Esperanto was officially born.

Zamenhof's utopian vision was of Esperanto as a new, neutral, non-imperialist world language which would overcome national enmities and inequalities. His first problem was that, if he wanted it to be a language which belonged to no one nation or ethnic group, what could he base it on? He himself spoke or knew Russian, German, Hebrew, Polish, Latin and Greek, but it would be hugely complicated to try to combine the grammatical rules and vocabularies of multiple languages. When he learned English at secondary school, he liked how simple the grammar was, so he started stripping away irregular verbs, exceptions and superfluous and little-used word forms from his language planning.

Then he turned to the vocabulary. First of all he tried to construct a logical system of words based on a, ab, ac, ad . . . ba, ca, da . . . aba, aca, ada . . . in which each element had a specific meaning. But he soon realized this was far too complex and almost impossible to memorize, so in the end he went to the Romance and Germanic languages, and tried to select the most internationally recognizable words. However, he was still swamped in massive vocabularies, so he decided on a system of root words plus affixes. In other words he could add invariable suffixes and/or prefixes to create words in the same semantic field without the necessity to learn each one separately. For example, the root-word *vend* (= related to selling) allows the formation of words like *vendi* (to sell), *vendejo* (store, shop), *vendisto* (salesperson, salesman), *vendistino* (salesperson, saleslady), *vendaĵo* (item for sale).

Based on the three principles of an international vocabulary, a regular grammar and word formation using affixes, Esperanto is a bit of a rag bag. Many root words look familiar to Europeans, while the diacritics give it a strange, exotic twist. For example: 'good morning' is *bonan matenon*, which seems pretty straightforward. But 'what is your name?' is *kiel vi nomiĝas?* 'Do you speak Esperanto?' is *ĉu vi parolas Esperanton?* And *kiom da blankaj birdoj flugas en la ĉielo?* means 'how many white birds are flying in the sky?' The word order is straightforward and noun and verb endings simple and consistent.

Today, Esperanto is used by anything between 10,000 and 2 million people worldwide. It's the most widely spoken constructed language in the world, and the only one which has native speakers, i.e., those who have learned from parents (such as the financier George Soros). It has been criticized for being too European-based in its vocabulary and grammar and therefore more difficult for, say, Asians to learn, but that hasn't stopped usage being particularly high in countries like China, Japan and Korea, as well as in Europe. Just as clearly, Esperanto hasn't realized its founder's vision of becoming a universal second language and bringing peace and unity among nations. It's too magnificent and impossible a dream. In that sense, Esperanto is the linguistic equivalent of a utopian world.

Zamenhof set out to construct a language which would be simple and easy, and various studies have suggested it can be learned in between a quarter and a twentieth of the time of other languages. If you surf the web, you'll find dozens of associations, clubs, articles, sites for learning and testimonials from Esperanto-speakers that learning Esperanto is easy and pain-free.

The popularity of Esperanto in Asia is curious, given its predominantly European vocabulary and grammar roots. Maybe it's the vision thing which appeals more, the idea of universal brotherhood and world peace. It's a kind of quasi-religious belief, which might explain why several small, non-traditional religions and sects have embraced Esperanto. Like Oomoto, for example, an offshoot of Shintoism, which was founded at the end of the nineteenth century.

Not only do most of its 45,000 adherents speak some Esperanto but they venerate Ludovic Zamenhof as a saint and visionary.

Other religions also espouse the Esperantist philosophy, including Bahá'í, which was the adopted faith of Zamenhof's youngest daughter, Lydia. Lydia was born in Warsaw in 1904, learned Esperanto from the age of nine, took a law degree and then travelled the world promoting Esperanto and her father's philosophy of brotherhood. She became a Bahá'í in the 1920s and continued to work, teach and promote both religion and language. While she was in Poland teaching Esperanto and translating Bahá'í scripture she was arrested by the Nazis because of her Jewish background and died in Treblinka in 1942.

Every single one of Zamenhof's children died in the Holocaust. His son, Adam, became an ophthalmologist like his father and was head of a hospital in Warsaw. He was arrested in 1939 and sent to Palmiry camp, where he was shot. His younger sister Sofia, a paediatrician, also died in Treblinka in 1942.

There are other names honoured in the Esperanto community. Petr Ginz was an immensely gifted Czech boy, a native speaker of Esperanto and a Jew. He wrote stories, drew and produced an Esperanto magazine called *Vedem* (We Lead). Before he was transported to Auschwitz, where he died in 1944, he passed on some of his writings and artwork to his sister, including a drawing called 'Moonscape' – a view of the earth from an imagined moon. Nearly seventy years later, in 2003, the first Israeli astronaut, Ilan Ramon, took a copy of that drawing on board the space shuttle *Columbia*. At the end of its sixteen-day mission, *Columbia* exploded on re-entry, and all seven astronauts were killed. A special stamp issued in memory of the shuttle featured a reproduction of Petr's sketch and a picture of him, young and smiling and happy.

Although the world does not speak Esperanto, millions of words of world literature have been translated into Esperanto, from the works of Shakespeare to the Koran. The Scottish Esperanto poet and translator William Auld translated Tolkien's *Lord of the Rings*. But it was for his own poetry in Esperanto that he was nominated for the Nobel Prize for Literature in 1999. There is a growing body of original literature by Esperanto writers all over the world, including the Icelandic novelist Baldur Ragnarsson, who learned Esperanto at school.

Esperanto has not become the world force its creator intended, but neither has it died out. Zamenhof, who died in 1917, is revered and celebrated by Esperantists all over the world, and there are two minor planets named in his honour – 1462 Zamenhof and 1421 Esperanto.

The language itself continued to change and develop after he died. There were schisms and arguments and reforms, resulting in offshoots like Ido. But Esperanto has held on to its place as the most successful constructed language, and it is part of the popular culture. So much so that, when Littlewoods ran a series of TV adverts for a new range of clothes in 2008, they decided to have the dialogue – between

An Esperanto version of Shakespeare's King Lear

mysterious, scantily clad women on a beach – in Esperanto. 'We believe it is a language that not only sounds beautiful, but exists to create harmony in the world, making this the perfect choice,' said a spokesman. Unfortunately, the Esperanto Association of Britain said it was really only cod Esperanto. But it's the thought that counts.

Globish: a Language to Do Business With

Globish is the name given to one of several simplified forms of English. It is not an artificial or constructed language, like Esperanto, but a fit-for-purpose, stripped down, simplified version of English put together for the business world. It is, claims the man who codified and now markets it, a form of global English, hence 'Globish'.

Jean-Paul Nerrière is a retired businessman, former naval commander and holder of the Légion d'Honneur. In the late 1980s he was a vice-president of IBM in America, in charge of international marketing. As he criss-crossed the globe, doing business with many different nationalities, he had to communicate in the one language which was common to all of them: English. Not perfect, literary English, but the get-you-by, pidgin English necessary for all non-native speakers wanting to do business in the global market.

The interesting thing Nerrière noticed was that the non-native speakers communicated with each other much more efficiently and easily than with native English-speakers. Whether it was a Japanese talking to a Korean, or a Moroccan talking to a Hungarian, the simple, basic form of English they used was mutually comprehensible and got the job done. On the other hand, the English used by native speakers, says Nerrière, is too complicated and idiomatic, and American or British businessmen didn't make the effort to simplify or adapt it in conversation. That could cost money. 'If you lose a contract to a Moroccan rival because you're speaking an English that no one apart from another Anglophone understands, then you've got a problem,' he says.

In essence, Nerrière's anecdotal and unscientific observations led him to this thesis: non-native speakers find a simple, common linguistic ground in order to do international business in English. He described it as 'the worldwide dialect of the third millennium'. What he did then was codify a Globish vocabulary of 1,500 words and publish it, along with a sort of mission statement, in his 2004 book *Parlez Globish*. He argued that non-native speakers had been finding their way round the subtleties and difficulties of English for years, by using simple words and describing things as best they could if they didn't know the idiomatic expression; in fact, doing what everyone does when they're speaking a foreign language. Nerrière simply took this a stage further by saying that there is no need to learn more than the basics required to be understood and to do business.

Nerrière identified the basics of English grammar, provided the cut-down vocabulary (most native speakers have about 7,500 words in their vocabulary) and developed rules and training to help non-native speakers use Globish as a lingua franca. He makes no bones about the utilitarian goal of Globish. It's 'a proletarian and popular idiom which does not aim at cultural understanding or at the acquisition of a talent enabling the speaker to shine at Hyde Park Corner', he wrote. 'It is designed for trivial efficiency, always, everywhere, with everyone,' he told *The Times* in 2006. It's designed for doing business.

It dispenses with idioms, literary language and complex grammar. It's no use for telling jokes – 'The only jokes which cross frontiers involve sex, race and religion and you should never mention those in an international meeting,' says Nerrière. So his books are about turning complicated English into useful English. For example, *chat* becomes *speak casually to each other* in Globish; and *kitchen* is the *room in which you cook your food*. *Siblings*, rather clumsily, are *the other children of my parents*. But *pizza* is still *pizza*, as it has an international currency, like *taxi* and *police*.

As befits a successful businessman, Nerrière identified his market, took out trade protection on the name 'Globish' and began to sell his books, manuals and training aids in shops, by mail order and on the internet. Interestingly, it's quite difficult to find many examples of what the courses consist of unless you buy them; they're not freely available by just surfing for 'Globish'.

Although Nerrière's claims for Globish are specific and utilitarian, this idea of a simplified, worldwide, cross-cultural adaptation of English is seen by some as a kind of third-millennium revolution where cultural barriers melt away, and the globe becomes one big Globish-speaking market place.

Globish may not be able to cope with jokes, but Nerrière still seems to be getting the biggest laugh out of his product. On the one hand he makes no great claims for Globish – it's not Shakespeare, it has no beauty or poetry, it just gets a very basic job done. But on the other, as a Frenchman who must have spoken English through gritted teeth in all his years at IBM, he can barely conceal his delight that Globish could perhaps be the canker that eats away at the dominance of proper, full-strength English. 'I am helping the rescue of French, and of all the languages that are threatened by English today but which will not be at all endangered by Globish. It is in the best interests of non-Anglophone countries to support Globish, especially if you like your culture and its language.'

A Klingon Hamlet

On a cold wintry day in a small theatre in Arlington, Virginia, Stephen Fry is appearing as Osric, the foppish courtier whom Hamlet mocks so accurately. But not any Osric. This version of *Hamlet* is in Klingon, the invented language that was created for one of the characters in the sci-fi TV series *Star Trek*. The Klingons (one planet, one people, one language, no Babel-like multiplicity of tongues) are those warlike, carapace-headed aliens who first appeared in a 1967 episode of *Star Trek* and proved so popular that they were upgraded from bit part to permanent crew in the later series and films. They fitted in neatly between the Vulcans and Romulans, and of course their distinctive foreheads made them instant party headpiece heroes.

Tonight's performance has been organized by a group of Klingon aficionados. Now a *Star Trek* convention is an event of myth and beauty, but an entirely Klingon gathering is a thing bizarre and ugly. The language is all guttural and rasping, and, befitting a warrior species, it's blunt, military and lacking in any delicacies or niceties. There's no word for *hello* in Klingon, just *what do you want?*

Stephen Fry with the creator of Klingon, Dr Mark Okrand

So how would the sublime poetry of *Hamlet* work in Klingon? Its inventor, linguist Mark Okrand, is here to unpack the mysteries of this fictional tongue. He explains that James Doohan, the actor who played Scotty, the Starship *Enterprise*'s engineer, made up the initial Klingon dialogue for the first *Star Trek* movie in 1979. This was dialogue in its loosest term – essentially an assortment of grunts. Okrand was asked to create a real Klingon language, with vocabulary and grammar, which actors could learn for later *Star Trek* movies.

'To make up a non-human language, first we have to know what the human language is, so we can make this one be "not that",' says Okrand. 'Human languages have certain things in common: they all have vowels, consonants, words, syntax. Certain sounds tend to go together in the same language and certain sounds tend not to. So, to make Klingon a non-human language, it made sense to violate the rules.' There were some constraints. The sounds had to be guttural, because the script said so. Okrand decided not to have a *z* sound in Klingon at all, 'because there's lots of zees and zots in science fiction.

Klingon grammar was made as non-human as possible. English, with its subject then verb then object, has the most common word-order structure: *Dog bites man*. Okrand chose the least common structure for Klingon with the object coming first then the verb and then the subject. So instead of saying, 'I boarded the *Enterprise*,' in Klingon you'd say 'The *Enterprise* boarded I.' Or, instead of Osric shouting, 'Look to the Queen there. Ho!' during the duel between Hamlet and Laertes, he cries (in Klingon) 'Ho! To the Queen there, look!' There are no tenses and no infinitives.

Klingon has a small vocabulary of around 2,000 words, many of them reflecting the bellicosity of the warrior race. Phrases like 'May you die well!' or 'Fire torpedos!' or 'Thrusters now!' are undoubtedly handy in galactic battle; not so helpful in day-to-day life. There are some wonderful insults too. 'Your mother has a flat forehead!' is one to be used with caution.

You would think that a language which has no word for love would make an uneasy bedfellow with Shakespearean prose. Why on earth would anyone choose to translate *Hamlet* into Klingon iambic pentameters? Well, in *Star Trek 3: The Search for Spock*, General Chang, Chief of Staff for Gorkon, the Chancellor of the Klingon Empire, has to quote some Shakespeare.

'We arrived on the set one day,' recalls Okrand 'and the director, Nick Meyer, says, "I need one more line: To be, or not to be." Now, one of the things about the grammar of Klingon is that there's no verb *to be*, so all I had was *or* and *not*. What I came up with was "Live or live not" which was *yIn pagh, yInbe*. I suggested it to Christopher Plummer, who was playing General Chang, and he goes, "YIn? YIn? That sounds too much like the Ying Tong song from *The Goon Show*. Think of something else." And I said, "Well, what about *taH pagh, taHbe*?' To go or not to go on?'

And so, in the movie, Chancellor Gorkon delivers the line 'You have not experienced Shakespeare until you have read him in the original and General Chang says: *taH pagh, taHbe*. To be or not to be, that is the question which preoccupies our people, Captain Kirk.' The Klingon Shakespeare Restoration project took up the Klingon challenge and has so far translated *Hamlet* and *Much Ado About Nothing*.

It's Mark Okrand and the Klingon fraternity's attention to detail which makes Klingon the most spoken, perhaps the boldest, of the invented fictional languages. It's difficult to learn, fiendish to pronounce, and yet people do seem to want to sit down and study and pass their Klingon certification exams. Some linguists find the academic potential irresistible.

Dr d'Armand Speers was a linguistics student at Georgetown University when he saw a flyer on the bulletin board advertising the Klingon Language Institute. He insists he wasn't a big *Star Trek* fan or into uniforms and prosthetic foreheads but he was intrigued by the advert. He joined up, learned to read and write Klingon and after some years became a fluent Klingon-speaker. In the meantime he got his Ph.D., married and had a child. And then he did something quite extraordinary. D'Armand takes up the tale.

'My son was born in 1994, and I decided as a linguistic fun thing to do that I would speak to him only in Klingon. I wanted to see if he would acquire the language in the same way that humans acquire human languages. And it was an interesting question because, when Mark Okrand created the language, he wanted to make it in ways that were different from human languages. So what would a human mind do when it encountered a language like this? There were two possibilities. The human mind would learn it as designed, meaning that we're more pliant than we think. Or it would change it into a more human structure. And either way it went, it would be a fascinating thing to see happen.'

D'Armand was the only one who spoke Klingon to the baby. The child's mother and everyone else spoke English. So what happened?

'Well, he was learning it,' says d'Armand. 'We had a lot of fun. We would play language games, and I would say things to him like "Where's my cheek?" in Klingon, and he would point to my cheek. And then one day we're playing, and I had his bottle that he would drink from, and we didn't have a word for bottle. We didn't have a word for diaper or high chair or most domestic things. I had words for shuttlecraft and phaser and transporter ionization unit but I didn't have bottle. So we were using the Klingon word for drinking vessel, *HIvje*, and I used it in a sentence. I didn't point at it, I didn't look at it, I didn't do anything like that, and this two-year-old toddler started crawling over towards the bottle and grabbed it. And at that moment I knew that this was working! He was learning this language, and it was very exciting.'

The language games continued. Every night father and son would cuddle up and sing a lullaby together – that soothing little ditty the Klingon Imperial Anthem, with the tender chorus 'Our empire is great, and if anyone disagrees, We will crush them beneath our boots.' The toddler began to learn to count in Klingon and was learning a few words and colours when suddenly, at around the age of two and a half, he stopped being interested.

'I would say something to him in Klingon and he would say it back in English,' remembers d'Armand, 'and I would try to encourage him, and he started to resist it. We would always count things, he'd count the stairs as he was walking up, and one day he was walking up the stairs counting in Klingon – "*wa', cha', wej*" – and then he looked and he realized that it wasn't me who was there but it was my wife, and he says, "four, five, six" and switches right in the middle of it.'

The Klingon conversations stopped. D'Armand's son had lost his motivation to acquire a language which only his father spoke. He wanted to communicate with his non-Klingon-speaking friends at playgroup and with his mother and other people around him. He gave up Klingon in the same way immigrant children tend to give up the parents' language once they're into their peer group.

D'Armand's son, a teenager now, is 'the sweetest kid' with no warrior tendencies or unfortunate guttural speech problems. He says he's keen to learn Klingon but, as he's passed the age of being able to acquire it naturally, he'll have to do it the hard way, learning it like any other second language.

Arika Okrent, the actress who played Gertrude in the Klingon *Hamlet*, is not only a level 1 Klingon-speaker and a fine Gertrude but she has a doctorate in linguistics, specializing in invented languages. She explains why she thinks Klingon is so popular.

'The reason this language is so successful and why fans do the work of learning it is because it does fit so well together in so many respects. And these days, if you want to create a fictional universe and make a film about it, it's no longer any good to just stick in some *zurg* and *blordon* and *zorbot*, you've got to make a fully realized, grammatical language, something that fans can sink their teeth into.'

Star Trek III: The Search For Spock, *1984*

The popularity of Na'vi, the language created for the film *Avatar*, is a similar case to Klingon. A fellow linguist, Dr Paul Frommer, undertook to create the language from scratch and used the tools he had learned as a linguist to make it work like any other language with its own syntactic rules. The success of *Avatar* has spawned a huge online Na'vi-learning community which now rivals Klingon to the extent that Dr Frommer is giving up his university job to concentrate on expanding the language.

An interesting question is whether the fact that the Na'vi in the film are rather gentle and mystical, quite unlike the Klingons, means that the people who speak it have to acquire some of their softer, nature-loving attitude. Arika thinks people choose a language in order to participate in a thing that suits them best.

But in serious linguistic terms has anything useful been learned by invented languages? Does it tell us about where language comes from and where it goes?

Arika has a theory: 'Well, it's given me a deeper appreciation for natural language and all its flaws and all its engineering problems and ambiguities and irregularities. They all serve a purpose and they suit the way that we think. And we in a way need our languages to be as messy and sloppy as they are in order to do the wide and wonderful range of things that we do with language. And if you want the whole range, you've got to have the sloppiness in there.'

Dialects and Accents

When is a language not a language? When it's a dialect, perhaps? There's no universally accepted answer. Some linguists say there's no difference at all. I suppose you could argue that French, Italian and Spanish are just 'dialects' of Latin. Or that Cockney or Geordie are languages. But that's just being difficult.

As we've seen, language is always evolving and changing, often without clear borders. The Yiddish linguist Max Weinreich cited this aphorism: 'A language is a dialect with an army and a navy.' That's clearly the case with, for example, Serbo-Croatian, the language spoken by the people of Serbia, Croatia, Bosnia and Montenegro when they were part of Yugoslavia. Since independence, however, each republic has established its own official language – Serbian, Croatian, Bosnian and Montenegrin. But they're all just dialects of the same South Slavic language. Contrast this with English, which is spoken in numerous places around the world but classified as different standard dialects – Standard American English, Standard Australian English and so on.

Max Weinreich's own Yiddish is classified as a language, even without an army, navy or nation. Occitan, Provençal, Breton, Welsh, Manx, Gaelic – they're all languages. So it's a subjective argument, but for the purposes of this chapter let's agree that a dialect is a variant of a language, usually regional. And in Britain we've got lots of them. From Scouse to West Country, Brummie to Cockney, Geordie to Scots, the British Isles are groaning with a rich and varied abundance of dialects. They're brimming over with words and expressions which take us right back to our linguistic roots – Anglo-Saxon, Latin, Norse, Celtic, Norman. Most dialects can be understood by non-dialect native English-speakers although accents and individual words can often flummox us.

Barnsley poet Ian McMillan lives and breathes Yorkshire and is a collector of local words, some of which can be pinned down to a single town or village. One word, *lenerky*, which means 'soft' or 'floppy', is found only in the tiny village of Grange Moor in West Yorkshire. There are some glorious words like *bartled*, to be smothered in something nasty, or *plotter* for mud or *gloppened* for astounded.

Ian says he speaks Barnsley rather than Yorkshire because within the dialect there are a huge range of accents, often changing from one village to another. He illustrates his point with a map of Yorkshire and shows how, in the space of just a few miles, the Yorkshire accent changes. His finger traces a journey from Hull in the east to Leeds ('long *es* and very slow') and Bradford ('they don't say their *ts*') and then down to Barnsley ('harsh . . . it's to do with the harsh winds of Yorkshire; you don't want to open your mouth too far'). Ian's finger hovers over the A61 as it makes its way down towards Sheffield. 'That's where an isogloss happens, where language changes over a small space. The *th* hardens to a *d*. So you start off from Barnsley, going "now then, now then", and as you approach Sheffield your vowels will be going "nar den, nar den". We call Sheffield people *deedars*, for the way they change the *th* in thee and thou to *d*.'

By the time Ian's finger stops its linguistic travels at the border with north Derbyshire, we've entered a foreign land. 'My Aunty Mabel, who was from Chesterfield, would say things like "I've just had double glazing fitted in my arse." Or she'd say, "I've got a detached arse."' Somewhere in the twenty-odd miles from Barnsley to Chesterfield, along the invisible isogloss line, *house* had changed to *hears* then to *harse* and ended up as *arse*.

Like all dialects, Yorkshire Tyke bears history. The traces of all the people who've come and gone and the work they did – it's all there in the layers of speech, says Ian, often without us knowing.

'Whenever I speak in my voice, behind me, in a huge line, are standing all the people who ever spoke like me. So that when I'm walking through Barnsley and somebody shouts to a lad, "Come on, clarteead, frame thissen!" you think, wow, that's a term from down the pit.'

Ian explains that a clout was a kind of nail used in the coal mines. *Clout head* or *clarteead* means 'idiot'. And *frame thissen* has Old English roots meaning 'get on with it!'

'So, somehow, the language carries on, using specific industrial terms that have gone, and they just hang on, like a kind of cloud or a ghost language . . . When the word dies, then the memory of a way of life will go as well. When a word isn't spoken any more then the roots of that word wither away.'

Ian remembers being told at school not to speak with his broad Yorkshire accent if he wanted to get on in life. 'And when I was first on the radio, a Barnsley listener rang up and said, "Tha' can't talk like that on't wireless." I said, "But I talk just like you." And he said, "Ar, but tha' can't talk like it on't wireless. They'll think tha're common!"'

A generation or so ago, speaking in anything other than Received Pronunciation – the Queen's English – was regarded as a social disadvantage. The ideal was to speak in a way that revealed nothing about your background, your class or where you'd been to school. As the elocutionist Arthur Burrell wrote in 1891: 'It is the business of educated people to speak so that no-one may be able to tell in what county their childhood was passed.'

RP developed from what was an essentially regional accent in the south-east of England. When William Caxton set up England's first printing press in 1473, he chose, largely, the spelling and pronunciation of the South-east/Midlands dialect as Standard English, as that was the most populous region of England and home to the royal court at London.

With the entrenching of the class system during the Industrial Revolution, the regional working-class accent increasingly branded the speaker as socially inferior. The development of public boarding schools like Eton and Rugby in the nineteenth century helped establish the RP accent as the 'standard' voice of the English gentleman, who went out into the world to build empires and occupied the highest positions in Victorian society.

The launch of the British Broadcasting Corporation in 1922 confirmed the place of Received Pronunciation as the accent of educated, middle-class England. Its first Director General, John Reith, said, 'One hears the most appalling travesties of vowel pronunciation. This is a matter in which broadcasting can be of immense assistance . . . We have made a special effort to secure in our various stations men who . . . can be relied upon to employ the correct pronunciation of the English tongue.'

RP became so synonymous with the dinner-jacket-wearing radio announcers that it came to be known as BBC English. This is what

Yorkshireman and BBC broadcaster, Wilfred Pickles

Wilfred Pickles rolls up his sleeves . . .

we would call today ultra posh, the sort of English spoken only by members of the royal family and the upper classes.

The BBC did in fact break with tradition during the Second World War, when it hired a Yorkshireman, Wilfred Pickles, as a radio news announcer. They thought that if the Nazis invaded Britain and infiltrated the BBC broadcasts, they'd find it more difficult to impersonate a regional accent. Radio announcers had already begun introducing themselves by name at the beginning of wartime news bulletins to ensure that listeners 'be able to recognize instantly the authentic voice of BBC broadcasting'. Wilfred Pickles' news broadcasts caused a furore, although if you listen to the BBC sound archives, it's rather hard to understand what all the fuss was about; you have to concentrate quite hard to distinguish the Yorkshire burr. Pickles did, however, infuriate southern listeners by ending the midnight bulletin with 'And to all in the North, good neet.' London cartoonists took their revenge by drawing him with rolled-up shirtsleeves and wearing a cloth cap.

U and Non-U

'It is impossible for an Englishman to open his mouth without making some other Englishman hate or despise him,' wrote Irishman George Bernard Shaw in the preface to *Pygmalion* in 1916. Forty years later, the English were still obsessing about the social classes and their accents. Linguist Professor Alan Ross coined the expression 'U and Non-U' in 1954 to differentiate between the English upper classes (U) and the aspiring middle classes (non-U). He argued that members of the upper class weren't necessarily better-educated, cleaner or richer than anyone else; the only thing which demarcated them was language (apart perhaps from 'having one's cards engraved ... not playing tennis in braces and ... dislike of certain comparatively modern inventions such as the telephone, the cinema and the wireless'). Do you *have* a bath (U) or *take* a bath (non-U)? Is your midday meal *lunch* (U) or *dinner* (non-U)? Do you wipe your mouth with a *table napkin* (U) or a *serviette* (non-U)? And do you use *lavatory paper* (U) or *toilet paper* (non-U)? Talking about *greens* instead of *vegetables* was very non-U, as was *home* instead of *house* and *glasses* rather than *spectacles*.

Ross's article was published in an obscure Finnish academic journal and might have passed unnoticed if it hadn't been read by the unabashedly U author Nancy Mitford. She immediately penned an article entitled 'The English Aristocracy', in which she expanded on the U and non-U theme and added a few of her own suggestions to a glossary of terms used by the upper classes. The essay provoked a debate about English class consciousness and snobbery. Mitford's article was reprinted along with contributions from Evelyn Waugh, John Betjeman

Queen Elizabeth II in 1952, making her first of many Christmas radio broadcasts

and others in *Noblesse Oblige*. Betjeman had a jolly good dig at the upwardly aspiring middle classes in his poem 'How to Get on in Society', which opens with the line 'Phone for the fish knives, Norman'.

Professor Ross asked the burning question 'Can a non-U-speaker become a U-speaker?' to which he answered no: 'in these matters, U-speakers have ears to hear so that one single pronunciation, word, or phrase will suffice to brand an apparent U-speaker as originally non-U'.

In fact, as Professor Ross, Nancy Mitford and Co. were discoursing social language, England's tight class system was already beginning to loosen. As the distinction blurred, so too did the boundaries between accents. What's interesting about the lists of U words is that today most of them have been adopted by everyone else.

The most famous purveyor of U-speak today is the Queen, but even Her Majesty's accent is showing signs of change. An Australian study of the Christmas broadcasts from the 1950s to the 1980s suggests that the Queen too is becoming less posh. Her vowels, apparently, have flattened slightly over the years, moving towards the modern standard accent of southern England. In the 1950s, her pronunciation of the word *had* almost rhymed with *bed*, and the words *pat*, *mat* and *man* sounded like *pet*, *met* and *men*. Today her accent is more like that of a posh Radio 4 announcer.

So if the accent of the U-class has become much more a standard RP, what of the non-U? Well, apart from RP and local accents, the accent for the non-U of the south-east of England is developing into a so-called Estuary English – somewhere between RP and Cockney. In EE, the *th* sound is becoming *f* or *v*; *t*s are dropping at the ends of words and the final *l*s in words are sounding like *w*s. EE, it's said, is less posh than RP but not as working-class as Cockney. In its strongest form it's the 'Am I bovvered?' accent of Catherine Tate's comic character or the 'yoof' drawl of journalist Janet Street Porter. Tony Blair is supposed to have adopted a mild form of Estuary English as prime minister when he wanted to appeal to Joe Public.

We're in a call centre in Newcastle, one of the boom industries of the twenty-first century. Over a million Britons are employed by companies as telephone agents to deal with customer inquiries or travel bookings or a myriad of other transactions. In this industry, regional accents – most of them, anyway – are prized possessions, and some are gold dust. Survey after survey puts Yorkshire and Geordie accents high on the warm, friendly and 'pleasing to the ear' scale. A Scottish accent gets top rating for most trustworthy; Welsh and Irish score highly as well. (Some accents don't fare so well: poor old Liverpudlian Scouse and Birmingham Brummie languish at the bottom as the most unappealing accents in Britain.)

The Newcastle call centre manager, Lawrence Fenley, explains that his employees are proud of their Geordie accent.

'Why should they hide it? Everybody in this centre is very proud of where they're from and their heritage. They will speak to the customers the way that they do to their friends, and it goes down very, very well.'

Clearly we humans have an emotional need to latch on to language; if you can't see the face of the person you're talking to, then the next best thing is their accent, because somehow we feel it's an expression of their character, their cultural belonging. Something like a Geordie accent has a very particular quality.

Brand consultant James Hammond makes this very point.

'In the twenty-first century things have changed an awful lot in terms of how we understand, appreciate and fit in with accents. And what we've got now is a society which is much more into an emotional connection. And, therefore, this Received Pronunciation, this Queen's English, is deemed almost condescending to us. We want accents that are much more personable . . . and we've got a celebrity status society now. So people like Ant and Dec and Cheryl Cole have really brought something like the Geordie accent to the fore.'

James does point out that RP is still the accent of choice when people want to speak to someone in authority.

'It is only when, usually, you have an issue, a problem, where that needs to be escalated up to somebody in a superior position, a manager or whatever, that you then expect a certain kind of accent. And, funnily enough, we then revert back to the Queen's English, because people feel more comfortable with that.'

Ant and Dec helping develop our love affair with the Geordie accent

Until relatively recently people thought having a posh accent was the key to success in life; that you'd be trusted and believed if you spoke like that. It's just not the case any more.

Yorkshireman Wilfred Pickles would have been delighted. As he wrote in his autobiography in 1949: 'May it be forbidden that we should ever speak like BBC announcers, for our rich contrast of voices is a local tapestry of great beauty and incalculable value, handed down to us by our forefathers.'

The last word must go to Barnsley poet Ian McMillan, a man who believes he is defined by his language and accent.

'Barnsley's what I think with. I think with its history, I think with its culture, I think with its hills that you walk up and get out of breath, I think with its wind that stops me talking in big words, big, big mouth openings. So yes, I think in the end, no matter how I try to write on the page, it will always come out with Barnsley'.

CHAPTER 3

Uses and Abuses

Language is the most versatile of tools, far exceeding the uses for which it was first designed. We humans probably began to talk to each other about the best places to hunt or which plants to avoid eating as a means of survival, but the way we use words has evolved into something much more complex. Our language can be playful and shocking and witty and dangerous. It allows us to express our anger verbally rather than physically, with expletives, and to profane the sacred or confront taboos; we hide behind the language of doublespeak to conceal truths, and cover our embarrassment with euphemisms; groups create secret codes for their protection or disguise or uniqueness; professions develop their own language for efficiency and accuracy; sometimes we use language simply to cover up the fact that we haven't a clue what we're talking about.

Like all complex tools, language can go wrong. It's when words erupt from us wildly, uncontrollably or don't come at all that we are given tiny insights into the mystery of how language works.

George Carlin backstage, March 1976, Pittsburgh

The Seven Words

Those of you with a sensitive disposition might like to skip the next few pages, for this is a chapter about bad language – whether it's jargon, slang, terms of abuse or, of course, swear words. If you choose to read on, be warned. The most offensive words in the English language are about to get an airing.

Let's cut to the chase and get the words down on the page. American comedian George Carlin was the first to list the Seven Words You Can Never Say on Television in a radio monologue in 1972. He reckoned they were: *shit, piss, fuck, cunt, cocksucker, motherfucker* and *tits*. 'Those are the heavy seven. Those are the ones that'll infect your soul, curve your spine and keep the country from winning the war.'

According to linguist Steven Pinker there are five swearing distinctions: dysphemistic (*I have to take a shit*); abusive (*Fuck you!*); idiomatic (*I was pretty fucked up last night*); emphatic (*I'm not going to do a fucking thing*); and cathartic (*Fuck, I've spilt my coffee!*).

The fact is, lots of us swear. It's become part of everyday life for more and more people, especially those under the age of thirty. People who count such things estimate we swear anything between fourteen and ninety times a day and half the times we use *fuck* and *shit*. So should we be bothered that it's now okay to repeat these most offensive of oaths on air and in print without retribution? Was Oscar Wilde right when he said, 'The expletive is a refuge of the semi-literate'? Or did Shakespeare get it right when he wrote: 'But words are words. I never did hear / That the bruised heart was pierced through the ear'? Is there such a thing as good or bad language, and is it advisable, or even possible, to control it?

Taboo or not taboo, that is the question.

Taboos

The word *taboo* derives from the Tongan word *tabu*, meaning 'set apart' or 'forbidden', and was first used in English back in 1777 by the explorer Captain James Cook. After observing the eating habits of the people of the South Pacific islands, he wrote:

> *Not one of them would sit down, or eat a bit of any thing . . . On expressing my surprise at this, they were all taboo, as they said; which word has a very comprehensive meaning; but, in general, signifies that a thing is forbidden . . . When any thing is forbidden to be eaten, or made use of, they say, that it is taboo.*

Hunters with their dogs corner a bear in its cave: the first taboo

And so *taboo* entered the English language, meaning something forbidden. Taboo subjects in almost all societies tend to involve religion, sex, death and bodily functions; things that frighten us or make us uneasy. The first taboo word in the Proto-Indo-European language was apparently for the animal we call the bear. It was so ferocious, so feared that people were scared to give it a name. It was referred to as the honey-eater or the licker or the brown one (*bruin* in Old English, from which we get bear).

Societies have developed a raft of words and expressions to deal with taboo subjects in contrasting ways: either to avoid mentioning the taboo subjects at all or deliberately to use the taboo words to inflame, to hurt, offend and shock. Sometimes it's simply for emotional release.

Take the subject of death. Given that it's the only certain thing in life (apart from, perhaps, taxes), we're remarkably coy about using the d-word. Instead, we say he's *passed away, passed on, given up the ghost, gone to meet his maker, shuffled off this mortal coil* or *joined the choir invisible*; we use euphemisms (from the Greek word meaning 'use of good words') to sweeten the harsh reality. If we want to laugh at death or be brazen about it, we use dysphemisms (Greek for 'non-word'). He's *croaked, pegged out, pushing up the daisies, bitten the dust, popped his clogs, cashed in his chips, kicked the bucket, called it quits* or (quite awful) *taking a dirt nap*.

We do the same for bodily functions. We *fart, break wind, pass gas, let rip, toot, poot, pop, parp* or *trumpet*; rather than defecate, we *poo* or *do a number two, have a shit* or a *dump, shed a load, test the plumbing* or – if you're in the medical profession – *have a bowel movement*. As for urinating, our replacement expressions are endless: we *pee, piddle, piss, slash, wee, widdle, have a jimmy riddle, go for a tinkle, take a leak, spend a penny, strain the potatoes, water the garden/tulips/tomatoes,* and, if you're a man, *point Percy at the porcelain* or *shake hands with an old friend*.

There's a lovely expression: to mince one's words, to *mince* meaning to soften or moderate. The way society has got round the problem of using outright profanities is to express a minced oath. We take a four-letter expletive, add a bit of rhyme or alliteration and come up with a softer, inoffensive version. *Flipping, frigging, fecking*, *sugar*, *shoot* and *shucks* – so innocent-sounding and yet so not. And if you thought you were being mild calling someone a *berk* instead of an idiot, think again: it's from the rhyming slang 'Berkeley Hunt'.

In our more pious past, words were altered to avoid blasphemy. Some of the euphemisms – *zounds* (God's wounds), *gadzooks* (meaning God's hooks, perhaps from the nails of the cross) and *God's bodkins* (God's body) – have fallen out of use. But our language today is still littered with expressions originally used to avoid taking God's name in vain. We say *good gracious, golly, gosh, by gum, begorrah, strewth* (God's truth) or *cor blimey* (God blind me). We make a detour round Christ or Jesus by uttering *cripes, crikey, for crying out loud* (for Christ's sake), *gee, jeepers* or *Jiminy Cricket* (did anyone tell Walt Disney?). Instead of hell and damn, we say *heck* and *darn* or *darnation, dang* or *doggone*. And *what the Dickens!* isn't alluding to the author but substituting the devil with a minced oath.

Why We Love to Swear

Professor Timothy Jay is a naughty words expert (what a great line to put in your CV). He's a psychologist at Massachusetts College of Liberal Arts and for the past thirty years he's been studying the role of dirty words in linguistics. Jay himself was exposed to swearing from an early age. His father was a carpenter, and the young Timothy would visit him at work on the building sites. 'I would hear things. The carpenters were planing a piece of wood, and the guy would say, "Take a few more cunt hairs off of that," or "I gotta cut that tit off there." So I would hear this language. If these guys came over to the house and they were with their families, you wouldn't hear any of it. But on the job – constant.'

Professor Jay calls swear words *emotional intensifiers*. 'It's like using the horn on your car, which can be used to signify a number of emotions,' he says. 'Like anger, frustration, joy or surprise.'

Swearing is so deeply rooted in the brain – both literally and metaphorically – that it stays with us longer. Jay claims that children learn swear words at a very early age and also learn the taboos that society places on them. 'As soon as kids can speak, they're using swear words,' says Jay. 'That doesn't mean they know what adults know, but they do repeat the words they hear.' His team's research revealed that children learn and use swear words as young as two or three. Because they're learned so early, they're deeply ingrained and like nursery rhymes are retained longer than other language forms.

Children learn that swear words are powerful; they often hear them during arguments when people's emotions are running high, and when the child repeats the word, they get an emotional response. It's the emotional link of swear words that gives them their psychological potency. Anything that is forbidden is powerful.

Jay sees an evolutionary advantage to swearing as well. 'We're the only animal that can express these emotions symbolically, so we can say "fuck you" instead of hitting you or biting you. Three-year-olds, before they really learn how to say "I hate you" or "fuck you", will bite you and scratch you. But when we learn how to use language to express that emotion, that primitive animal anger goes away.'

The idea that swearing is both evolutionarily and developmentally primitive is also being explored by linguist Steven Pinker. He argues that swearing and linguistic taboos tap into the workings of the deepest, most ancient part of our brains. He likens these deep brain responses to that of, say, a dog's brain: bumping your head and yelling out an expletive is the same as a dog yelping suddenly when you step on its tail – a sort of canine curse. 'Cathartic swearing,' he says, 'comes from a primal rage circuit, in which an animal that is frustrated, confined or hurt erupts in a furious struggle accompanied by an angry noise, presumably to startle and intimidate an attacker. Some neuroscientists have even revived Darwin's suggestion that verbalised outbursts were the evolutionary missing link between primate calls and human languages.'

Coprolalia

Clearly many of us like to swear, even if we disguise it as a minced oath. But there are some people who have no choice in the matter – they have to swear. Uncontrollable, often foul words erupt from them; the urge to speak the unspeakable impossible to deny. They have a condition called coprolalia (from the Greek *kopros* – faeces – and *lalia* – talk), a disorder which we probably associate most with Tourette's syndrome – although only around 15 per cent of Tourette's sufferers have involuntary coprolalia.

The inappropriate language seems to vary from culture to culture, depending on the different taboos. So in Catholic Brazil you'll get references to the Holy Mother; in Asian cultures, where the family is honoured, you get outbursts like 'shitty grandma' or 'aunt fucker'. Most English-speaking Touretter's say the same sort of things – 'fuck, shit, hell, cunt' – and they don't usually use euphemisms. It's the socially inappropriate words that they can't put a brake on.

Jess Thom has suffered physical and verbal 'ticks' since she was a small girl. The physical ticks are unpredictable and exhausting; the verbal ones involuntary and socially difficult. It's a bit like having a jack-in-the-box in the brain, on such a strong spring that it's forever breaking its latch and popping out at the most inopportune moments. Jess goes through phases with words – swear ones as well as random day-to-day ones. Today her recurring involuntary word – apart from 'fuck' – is 'biscuit', with the occasional 'Happy Christmas' thrown in. She describes her earliest memories of having Tourette's.

'I had noises. The first fuck noise I can remember was a squeaky one when I was about six. My ticks when I was younger and all through my childhood were much more motor and also much more mild fuck than they are now. Fuck. For lots of people Tourette's gets better as they get older. Fuck. For me in adulthood and in my early twenties my ticks got much more noticeable to other people fuck, although the sensation for me biscuit didn't change that much. Fuck.'

Jess's sentence construction is almost perfect – despite the constant involuntary interruptions. Listening to her, it's clear that the considered sentence is coming from one part of her brain and the 'biscuits' and 'fucks' from another part. It's as if there's another person in the room, butting in on the conversation.

'It's going all the time biscuit, and my thoughts are clear. It doesn't very often interrupt my thoughts . . . Sometimes my sort of thinking fuck sometimes it does. Sometimes I'll be put off my train of thought by ticks. But very unusual though. Ha biscuit. Mostly I sort of know what I'm saying fuck fuck. What I don't know that I'm saying, what I'm not choosing to say, is all the fuck, all the ticks which just sort of frustratingly interrupt. They're not communicative. It's sort of eighty per cent biscuit of the language I use doesn't have a communicative purpose or intent. Fuck. Happy Christmas.'

Jess's family and friends have learned to pick their way through her sentences, zoning out the random words. She says they're even able to distinguish an intended expletive from an involuntary one.

'I was speaking to my dad on the phone fuck the other day, and he's used to very rude swearing in our conversations constantly peppered with ticks but sort of understands them for what they are. Fuck. But then I used fuck to describe something. I said something was fucking something and he knew instantly and told me off and told me to mind my language. Fuck. And it really made me laugh 'cos it was like he hadn't heard all the offensive words because he knew they were ticks and had no meaning, but as soon as I'd used something deliberately, he pulled me up on it.'

Proof, if ever it was needed, that it's not the words themselves that matter but where they come from.

'Absolutely,' agrees Jess. 'I think lots of people misunderstand Tourette's and say, "I wish I had Tourette's, it could mean I could get away with swearing or it means I could say whatever I biscuit biscuit I could say whatever I wanted to." The whole point is I can't say whatever I want to. Lots of what I say I don't want to say. It's just there fuck and it's biscuit biscuit biscuit Happy Christmas, but you know that doesn't mean that I can't articulate my thoughts and make myself understood. Fuck. Biscuit.'

Deaf Tourette's

A fascinating addendum to verbal ticks is that there have been cases of deaf Tourette's patients who swear compulsively in sign language. The medical journal *Movement Disorders* reported a case study from 2001 of a thirty-one-year-old man who was deaf from birth and who had motor and vocal tics as well as coprolalia.

> *He would feel a compulsion to use the sign for 'cunt' (see Fig. 1) in contexts (grammatical and social) that were not appropriate. This is essentially the sign for the medical term 'vagina' except that the sign is pushed toward the person at whom it is aimed and accompanied by threatening body language and facial expression. The patient would then feel embarrassed about the compulsion and aim to disguise it as another sign. Commonly, this would be the sign for 'petrol pump' (see Fig. 2). This can also be used to symbolise a small watering can.*

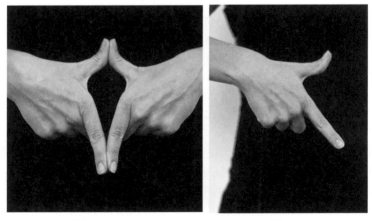

Fig. 1   Fig. 2

This single case illustrates clearly that coprolalia in deaf and hearing people with GTS [Tourette's] is not just a random utterance or gesture but one that conveys meaning and purpose.

The Mechanics of Swearing

The study of Jess and other Tourette's patients is providing fascinating insights for neurologists trying to identify the mechanics of why we curse. Time for a bit of science.

Around the edge of the brain's cortex is the limbic system, a complex network of deep brain structures which is thought to control our motivation and our emotions. Interestingly the same structures can be found in the brains of evolutionary ancient animals like the alligator.

On the outer layer of the brain is the neocortex, made up of folds of grey matter, which is responsible for higher functions like knowledge, conscious thought and reasoning. We process most of our language in the neocortex, but there are some words – curses and taboo words – which have strong emotional connotations (Timothy Jay's *emotional intensifiers*), and these are processed in a part of the limbic system called the amygdale. The amygdale is an almond-shaped mass of neurons at the front of the temporal lobe of the brain that appears to gives our memories emotion. Studies have shown that, if the amygdale is stimulated electrically, animals respond with aggression. If it's removed, the animals become tamer and don't react to things that would normally have angered or frightened them. In humans, brain scans show the amygdale light up, i.e. become active, when the person is shown a card with an unpleasant word, especially a taboo word, written on it.

'The response is not only emotional but involuntary,' writes Steven Pinker in his article 'Why We Curse. What the F***?' 'It's not just that we don't have earlids to shut out unwanted sounds. Once a word is seen or heard, we are incapable of treating it as a squiggle or noise; we reflexively look it up in memory and respond to its meaning, including its emotional colouring.'

As Timothy Jay argues, once we've seen or heard these emotional intensifiers, we can't erase them. And when things go wrong in the left hemisphere of the brain, where we process language, it's those emotional words lodged in the deeper limbic system which we are still able to access.

Possessed by the Devil

The French poet Charles Baudelaire was visiting the ornate carved confessionals of a Belgian church when he felt dizzy, staggered and fell. By the time he had reached his carriage, the forty-five-year-old poet's language had become confused – he asked for the window to be opened when he meant shut. 'One of mankind's greatest-ever language centres had started to die, for ever,' as Peter Silverton put it in *Filthy English*. Baudelaire's rich linguistic repository was cleared out. Within the month he could not speak at all – apart from one expression which, to the horror of the nuns who were looking after him, was the blasphemous curse *cré nom* (*sacré nom de Dieu*) – damn! The nuns thought Baudelaire had been possessed by the devil.

One hundred and fifty years later we know that Baudelaire had suffered a stroke. The blood flow to his brain had been cut off, causing damage to the language centres in the left hemisphere. This impairment of speech after a stroke is a condition known as aphasia. And the shouting of a single blasphemy, *cré nom*? Well, around 15 per cent of people who have aphasia have verbal automatisms, brief unconscious utterings which are often, although not always, swear words.

Les Duhigg suffered a stroke fourteen years ago, aged forty-one. When he came round in hospital, the first thing he said was, 'Where am I?' Except that he didn't actually say it. As Les recounts: 'In my mind I'd say that, but nothing could come out and I was . . . dumbfounded, not able to speak for the first time in my life.'

He was, literally, dumbfounded. And then, a few days later, he uttered his first word, 'FUCK!', when he heard the doctors discussing moving him to a side ward in which another stroke victim had just died. It gave the doctors – and Les – quite a shock. Les hadn't been much of a swearer before his stroke. It wasn't that he was now putting swearing into normal speech but that swearing was the only word he could generate. It wasn't 'Pass the fucking cup of tea', it was simply 'fuck'.

For a while he used the f-word for everything, much to the embarrassment of his wife, Marion. 'I was always apologizing for him, especially to the physiotherapist. But she'd worked with stroke people before and said it was a common thing.'

Les was unconsciously pulling out those emotional memorized words from the undamaged right side of his brain. He and Marion saw something similar happening with other patients.

Marion remembers: 'We had a chap in the stroke group. He couldn't speak, but if someone started singing a song and he knew it, he'd just join in with them. And we couldn't make it out because he couldn't speak properly and then all of a sudden he'd come out with singing.'

Singing songs learned as a child – like counting and nursery rhymes and swearing – is often automatic, and the patient can still produce the motor movements associated with a sequence, even though afterwards they can't retrieve the appropriate words to say what it was they were singing.

Les had to learn to speak all over again. The involuntary cursing has stopped and, fourteen years later, speech therapy is helping his brain learn new pathways to reinvent language. 'It's like being born again,' he says 'starting off as a little kid.'

Les is a patient of Professor Cathy Price at University College London. To help explain what is happening in the brains of aphasia sufferers like Les she suggests performing an experiment on a guinea pig, Mr Stephen Fry. She wants to show that Les's involuntary swearing wasn't a grasping on to the only emotional, automatic words he could access; rather, the mechanism which allows people to inhibit words and actions in social situations had been damaged

Cathy's plan is to wheel Stephen into an MRI scanner and record his brain patterns while he performs a *Just a Minute* routine – exactly like the long-running panel show on BBC radio, in which contributors have to speak for a minute on any given subject without hesitation, deviation or repetition. He is asked first of all to speak freely on various subjects that force him to use the left side of his brain, where information is stored. Sure enough, the scan shows the frontal lobes in the left hemisphere activating. Then he is given a variety of subjects to speak about without repetition, interspersed with bouts of counting. These scans are fascinatingly different. Although Stephen is still using the frontal lobes for the factual knowledge, a tiny little structure deep in the grey matter of his brain is flashing away. Cathy explains that this is called the left head of cordate within the basal ganglia structure, and one of its functions is as an inhibitor. The scans show it working hard as Stephen tries not to repeat his 'taboo' *Just a Minute* words – exactly as it does when one tries not to swear in front of children and old ladies.

Cathy compares this with what happens when someone who is bilingual is speaking – they have to focus on one language while suppressing the other one. Then she contrasts the MRI scan of Stephen's brain during the *Just a Minute* routine with Les Duhigg's brain scan. While the left head of cordate in the basal ganglia area – the inhibitor – lit up in Stephen's scan, in Les's it remained dark, damaged irreversibly during his stroke.

Brian Blessed's Swear Box

Actor Brian Blessed is a prolific swearer. Turning the air blue comes as naturally to him as breathing; he makes Gordon Ramsay seem like a choir boy.

Brian agrees to join Stephen in an experiment conducted by Dr Richard Stephens, one of whose specialities is the connection between swearing and pain. He devised this test after noticing how his wife seemed to get natural pain relief from swearing during childbirth.

In the middle of the room is a fish tank, filled with ice. The point of the experiment is to plunge a hand into the icy water and see how long they can keep it in. The first time they are only allowed to repeat a single word that could be used to describe a table; Stephen's is 'functional'. The experiment begins:

Richard: So, Stephen, when you put your hand in the water, I'd like you to repeat that word at an even, steady pace. Keep your hand in as long as you can and take it out when you're ready.

Stephen: That is cold actually. Functional. Functional.

Functional. It's beginning to hurt. Functional. Functional. It begins with the right leg. Functional. Oh fuck.

Richard: Don't swear.

Stephen: I'm not to swear, I'm sorry. Functional. Functional. Functional. This really hurts. Oh this is not funny any more. Functional. I'm going to get hypothermia. Functional. Oh God, I can't take it, I'm sorry.

Richard: Right. OK I'd like you to do that again. This time I'd like you to tell me a word you might say if you hit your finger with a hammer.

Stephen: Well, I'm afraid I'll be dull and it would be 'fuck'. That would be the first one that would come, and then the many others would stream afterwards . . . And here we go. Oh yes. Good. Fuck. Ah ha. It's all right for the moment.

Brian: Is it worse this time?

Stephen: It's still cold and my hand . . . Oh fuck me. Oh fuck this for a game of fucking soldiers. Fuck. Fuck the fuck.

Brian: Terrible language.

Stephen: I'm so fucking sorry. Fuck it.

Brian: This is going to go all over the world. You're going to lose your reputation as an elegant person.

Stephen: Oh fuck, fuck, fuckedy fuck. Ooh. It feels better actually saying fuck. It actually doesn't feel so bad. Fuckedy. Ooh. Ooh. Very tingly. I think I'm ready to bring it out at any fucking point now but . . . you know I can keep it in here in a way that I couldn't before. I genuinely mean that. That's quite extraordinary. It just lets you. It does, doesn't it? I think I'm ready to take it out now.

Richard: Brian, you've seen the procedure; we'll do the same thing again. So we'll start with the word that you might use to describe a table.

Brian: Wooden.

Richard: Wooden. That's a good choice.

Stephen: And no swearing. No swearing.

Brian: Right. Oh, it's lovely and warm. Wooden. Wooden. Wooden. It is cold, isn't it? Mustn't swear. Oh, wooden. This is horrible, isn't it? Wooden. Wooden. Wooden. Oh, fuck.

Stephen: No.

Brian: Oh no, no wooden . . . Wooden. Wooden. Wooden . . . I'll take it out.

Richard: OK, Brian, and so this time I'd like you to use a word that you might use if you hit yourself on the thumb with a hammer. Can you give me your word that you might use?

Brian: Yes, I'd say bollocks. Fuck it.

Richard: Just one word.

Brian: Bollocks . . . Here we go. I always get terrible fucking wind. I'll be all right in a minute. I don't know why the fuck I do that. I get terrible wind. Here we go. Right. Oh bollocks. Oh bollocks. Bollocks. Bollocks. Bollocks. Oh bollocks. Is that all I can say is bollocks?

Richard: Steady, even pace please.

Brian: A steady even . . . Fucking hell, man. Bollocks. Bollocks. Bollocks. Bollocks. Oh fuck it.

Richard: That's great. Thank you . . . This couldn't have really gone any better.

The results of the experiment are revealing. Stephen is not an inveterate swearer, so, like the majority of people who have taken part in the test, he tolerated pain better when he swore. He kept his hand in the icy water for thirty-eight seconds with his neutral word but for two minutes and twenty-nine seconds with his swear word. Brian, on the other hand, is an habitual swearer, so swearing appeared to have no effect on his pain threshold at all. In fact, he kept his hand in for five seconds longer when he shouted 'wooden' than when he swore.

TV Humour and the Curse

When uttered at the right moment, a rude word can suddenly bring an otherwise dull and lifeless sentence dramatically to life. In the rather dry world of humour research, this is known as a 'jab line'. It adds emphasis and a touch of the unexpected, a necessary component of humour. It is often particularly funny when coming from an unlikely source, such as the mouth of a sweet old lady or a seemingly innocent child.

One of the best-loved comedy sketches on British television is *The Two Ronnies* 'The Swear Box', a masterpiece of innuendo, in which the anticipated expletives from two men in a pub were bleeped out by a volley of increasingly strident beeps. That was in 1980, when swearing on television was still uncommon. There had been a sprinkling of 'bloodies' and 'damns', including the forty-four 'bloodies' repeated in an episode of *Til Death Us Do Part* in 1967, after which the broadcast standards campaigner Mary Whitehouse declared 'This is the end of civilization as we know it.' And, of course, there was the famous late-night 'fuck' uttered by Kenneth Tynan on live TV two years earlier, which caused a national uproar and prompted one Tory MP to suggest Tynan should be hanged.

Today the use of expletives on television after the 9 p.m. watershed is widespread. Nowhere has swearing been taken to such operatic levels as in the BBC's satirical political sitcom *The Thick of It*, with its foul-mouthed Downing Street spin-doctor Malcolm Tucker. *The Thick of It* has been described as the twenty-first century's answer to *Yes, Minister*, the gentler but equally witty sitcom of the 1980s. Both programmes satirized the inner workings of British government; their language is very different. 'Gibbering idiot' is about the most

Mary Whitehouse, campaigner for broadcast standards

extreme form of abuse used by hapless MP Jim Hacker in *Yes, Minster*, whereas 'Please could you take this note, ram it up his hairy inbox and pin it to his fucking prostate' is a typical 'Tuckerism' from *The Thick of It*.

Armando Iannucci is the show's creator, writer and producer. He says he is simply reflecting the language of the government's inner circle in the first decade of the twenty-first century.

'There was that world which lived off a twenty-four-hour news cycle, it lived off a "we've got to control every media outlet possible" and therefore every second was a battle, which is why the language started getting more hot-tempered. But it's different for different factions. I've done a bit of swearing research and [Prime Minister] Cameron's troops don't swear as much as Gordon Brown's troops. When we were doing *In the Loop* [the film version shot in USA], I established that the State Department didn't really swear that much, but the Pentagon swore like dockers, they were absolutely filthy, so we injected that into it. So it's really there to reflect the reality . . . There is something enjoyably childish about it . . . it does feel like

Smooth-talking Permenant Secretary Sir Humphrey, Private Secretary Bernard and hapless Jim Hacker

you're breaking a rule somewhere, but nobody's dying as a result, you're not causing any physical harm.'

Jesuit-educated Iannucci shares the concerns of those who worry that the overuse of expletives devalues the language of comedy.

'The last thing I want is every programme I watch to be like that; that would be boring apart from anything else . . . I'm not a swearer although I do find swearing funny; I see the funny side of it but I do find it quite tiring. If I'm watching a stand-up who is just f-ing and blinding every other one, I find it a little bit dull because it just becomes sort of incessant and numbing . . . so I like the creative use of swearing.'

The Thick of It, *with spin doctor Malcolm Tucker, played by Peter Capaldi, and its creator, Armando Iannucci*

Euphemisms

If cathartic swearing – the expletive shouted when you stub your toe – is primal, then what about the rest of our taboo language, the euphemisms and minced oaths, the carefully crafted word used to replace the blaspheme or the improper?

There's a school of thought which says that it was our very need to find ways of avoiding taboo subjects which propelled humans into developing complex language. 'Euphemism is such a pervasive human phenomenon,' wrote linguist Joseph Williams, 'so deeply woven into virtually every known culture, that one is tempted to claim that every human has been pre-programmed to find ways to talk around tabooed subjects.'

Euphemisms have been described as a barometer of changing attitudes. In more religious times, the need to avoid open blasphemes was strong, and yet subjects like sex and bodily functions were often not taboo at all. Look at Chaucer's language – full of vulgar, bawdy expressions. And the earthy richness of Shakespeare. His patron, Queen Elizabeth I, was said to swear like a man and enjoy dirty jokes. In his *Brief Lives*, the seventeenth-century diarist John Aubrey recounted the story of Edward de Vere and his unfortunate deep bow to the Queen.

> *This Earl of Oxford making his low obeisance to Queen Elizabeth, happened to let a Fart, at which he was so abashed and ashamed that he went to Travel, seven years. On his return the Queen welcomed him home, and said, 'My Lord, I had forgot the Fart.'*

Restrictions on blasphemous language reached a peak under the Puritans. Oliver Cromwell warned his armies that 'Not a man swears but pays his twelve pence.' A quartermaster called Boutholmey was condemned to have his tongue bored with a red-hot iron for uttering impieties; one William Harding, of Chittlehampton, was found guilty for saying 'upon my life', and Thomas Buttand was fined for exclaiming 'on my troth!' The Puritans who left England to settle in

Edward de Vere, Earl of Oxford, made a rather embarrassing faux pas in front of his Queen

America took with them similar strictures. A blasphemer could be put in the pillory, whipped or have his tongue bored out with a hot iron.

As society became less religious, it began to lose its horror of impious language. But one fear was soon matched and then overtaken by another – that of impropriety. Words to do with body parts, body functions and especially anything to do with sex became increasingly euphemized. The American writer H. L. Mencken dubbed the early nineteenth century a 'Golden Age of Euphemism' – on both sides of the Atlantic. The middle classes in particular went to extraordinary lengths to 'purify' their language so as to avoid the slightest chance of an improper thought.

This was the era of the prude, when *leg* became *limb*, *breast* became *bosom* and *belly* went from *stomach* (an old Latin word) to *tummy* or *midriff*. Ladies didn't *sweat*, they *perspired*; they weren't *pregnant* but *in a delicate condition*; and people didn't *go to bed*, they *retired*. *Trousers* were called *unmentionables* or *inexpressibles*; *underwear* was *linen*.

The Americans seemed to be particularly squeamish, changing *haycock* to *haystack* and *weather cock* to *weather vane*. In the farm-yard, *cockerels* were now *roosters* and *bulls* – symbols of sexually potency – were *male cows*, *gentlemen cows* or *cow brutes*. Even the English word *titbit* became *tidbit*.

The author of *Little Women*, Louisa May Alcott, was the descend-ant of settlers who arrived in America in 1635 with the surname Alcock. Through the next few generations, as embarrassment over the improper connotations of the name grew, the family changed it to Alcocke and then Alcox. Louisa May's father went the whole hog in the 1820s and changed his name to Alcott.

Even a cooked chicken on the dinner plate didn't escape the prudes. A chicken *drumstick* replaced the simple chicken *leg*. Chicken *breast* was, of course, far too rude to say, so it became *white meat*, while the sexy *thigh* transformed into *dark meat*.

Winston Churchill fell victim to southern American modesty when he attended a dinner in Richmond, Virginia. The butler came round with a plate of chicken and asked Churchill which piece of the bird he'd like, to which Churchill replied, 'I'd like breast.' The hostess sitting next to him blanched and said, 'Mr Churchill, in this country we say white meat or dark meat.' The next day he sent her a corsage of flowers with the message 'I would be most obliged if you would pin this on your "white meat".'

Medicine and Euphemism

Medicine and euphemisms have long been bedfellows. For centuries doctors and nurses have referred to parts of the body using Latin technical terms that are themselves euphemisms borrowed from another culture. The word *penis* is actually a Latin word meaning 'tail', and *vagina* is a Roman synonym for 'sheath' or 'scabbard'. Doctors insist that the Latin terms are necessary for precision. So they talk of *mammary* (breast), *cranium* (head), *metacarpal* (wrist) and *phalanges* (fingers). A British soldier shot in the buttocks during the First World War was asked by a visitor where he'd been wounded. 'I can't say,' he replied. 'I never studied Latin.'

Diseases have always been euphemized – indeed the word itself, *dis-ease*, is a gentler substitute for sickness. Consumption sounded much more romantic, if no less deadly, than tuberculosis. Early terms for syphilis deflected fear by insulting the enemy instead. To the English it was *Spanish pox*, *Neapolitan bone ache* or *malady of France*. The Poles called it *German disease*, whilst the Russians called it *Polish disease*. The Turks preferred *Christian disease*. And it's only recently that people have felt comfortable saying the word *cancer* instead of *Big C*. Or not mentioning it at all.

One of the most euphemized places is a hospital. We get terribly embarrassed talking to strangers about what our bodies should be doing quite naturally; instead we prefer to say things like *private parts* or *down there* or *waterworks* or, heaven help us, *the doings*. This can cause all sorts of problems for the nursing staff, especially for foreign nurses.

The Queen Elizabeth Hospital in King's Lynn, Norfolk, runs a course to help disambiguate its new staff. Today's trainee nurses are all Portuguese with excellent English. Perfect English may not be enough, warns Staff Nurse Julia Saunders, and she illustrates the problems of our euphemism-laden language with stories from the ward.

'We had an auxiliary who was Portuguese. He was on the ward one day, and a lady called him over and says, "I need to spend a penny." And he said, "That's fine, I'll be with you in a moment, I'll just finish what I'm doing." So she again called him over and said, "I need to spend a penny," and he said, "I truly will be with you, madam, in a moment," being very polite, and then the third time he went over and said, "My dear, the paper lady's in the next bay and you can spend as many pennies as you like when she comes." Then the Staff Nurse came in and said, "George, she needs to go to the toilet." And he was mortified, he said he felt so silly, he really didn't understand, he wouldn't have made her wait if he'd realized what that phrase meant.'

'People will ask for a bottle, and I've had people running around giving them bottles of lemonade, bottles of juice, bottles of water and actually what they want is a urinal, but it's the common terminology – a gentleman will ask for a bottle.

'"Rose Cottage" is a terminology that we use for the morgue or the mortuary here. It's kind of going out of fashion, but it's a word that's used throughout the NHS, and people tend to think it's a nicer terminology. If you're standing at the desk and you ring the porter and say, "I have a gentleman for Rose Cottage," well, that tends to sound better than "One for the morgue", doesn't it? And in Paediatrics, they sometimes say, "I've got a little one for the Rainbow's End."

'I've had a lady come in and she's said, "You've lost him, what d'you mean you've lost him, have you got a search party out, where is he?" And we're going, "No, no, no." And the nurse was getting herself more and more in a pickle, simply because she thought she'd know what she meant. So I said, "I need you to sit down," and I said, "Your husband has actually died." And when I said that word, although it was very harsh perhaps, she actually understood what I meant.

'I had a reasonably junior doctor and I think it was probably his first time breaking some bad news to the patient. He sat down with the lady, and I was there obviously to comfort. I knew what the bad news was. And he told her that she'd got a malignant tumour, and I remember, as a young nurse, thinking, oh, this lady's taking it very well. She asked a couple of questions about treatment, and he said there's not really any at this stage (I'm talking twenty-five years ago), and he quickly left the room, and I thought, gosh, if I'd been told that news and I was only thirty or thirty-five, I don't think I'd be sitting there like she is. So I said, "Did you understand what the doctor said?" She said, "Oh, d'you know, I was really worried when he brought me in here that he was going to tell me I'd got cancer." And I said, "Well, what do you think?" "Well, I've only got a malignant tumour." And I suddenly realized that the key word for her was *cancer*. Because cancer hadn't been heard . . . that's what she needed, she needed to hear that exact word.'

Innuendo

Euphemisms, innuendo and double entendres have long been a mainstay of British humour. In the world of entertainment, they've allowed performers to keep their acts clean enough to escape censorship and everyone in the family to enjoy the comedy. While younger members take the statement at face value, older members can enjoy the more risqué meanings. Be it a Golden Classic – 'A woman walks into a bar and asks for a double entendre, so the barman gives her one' – or the unintentional bloomer – 'Ah, isn't that nice. The wife of the Cambridge president is kissing the cox of the Oxford crew' (commentator Harry Carpenter at the 1977 Oxford–Cambridge boat race), innuendo does seem to be a particularly British obsession.

Shakespeare frequently used innuendos in his plays. Hamlet taunts Ophelia with sexual puns, referring to 'country matters' and Sir Toby in *Twelfth Night*, describing Sir Andrew's hair, says 'it hangs like flax on a distaff; and I hope to see a housewife take thee between her legs and spin it off'.

From the mid nineteenth century onwards, the music hall kept innuendo alive through an age of Victorian prudery. Queen of the music hall and the double entendre was Marie Lloyd, whose delivery of a song or a line ('She'd never had her ticket punched before') was accompanied by saucy winks and gestures. If her trademark parasol failed to open, she'd quip, 'I haven't had it up for ages.' Lloyd locked horns with a Mrs Ormiston Chant of the Purity Party, who made a public protest against her from the stalls of the Empire music hall in London's Leicester Square. In 1896, Lloyd was summoned to appear before the Vigilance Committee so that it could decide whether her songs were a threat to public morality. She sang two of her most famous songs – 'Oh Mr Porter' and 'A Little of What You Fancy' – without her usual winks and gestures, and the committee had to acquit her. The story goes that, after the demure performances, Lloyd stunned the room with a rendition of 'Come into the Garden, Maud', accompanied by an array of obscene gestures. Another story has Marie getting into trouble with her song

Marie Lloyd accompanied her songs with saucy winks and gestures

'She Sits among the Cabbages and Peas'. She continued to sing it but merely changed the lyric to 'She sits among the cabbages and leeks.'

On a visit to the United States, Lloyd explained her style in an interview with the *New York Telegraph*: 'They don't pay their sixpences and shillings at a music hall to hear the Salvation Army. If I was to try to sing highly moral songs, they would fire ginger beer bottles and beer mugs at me. I can't help it if people want to turn and twist my meanings.'

More than 100,000 people attended Lloyd's funeral in London. One of her fans, the poet T. S. Eliot, described her death as 'a significant moment in English history'. As London correspondent of the *Dial* magazine, he wrote: 'No other comedian succeeded so well in giving expression to the life of the music hall audience, raising it to a kind of art. It was, I think, this capacity for expressing the soul of the people that made Marie Lloyd unique.'

Cheekie Chappie Max Miller dominated the music halls from the 1930s to the 1950s, at a time when the office of the Lord Chamberlain was busy censoring plays and scripts for lewdness. Max Miller never swore on stage or told a dirty joke but he took innuendo to new heights of vulgarity. He got round the censors by carrying two pocket books on stage with him, one white and one blue. He'd explain to the audience that they were joke books and asked them to choose which one they'd like. If they chose the blue one, the one with all the risqué jokes, it was their own choice. He'd look off stage as if checking to see if the manager was there today, then he'd beckon to the audience and get on with the rude stuff.

Miller often used 'mind rhymes', which left the audiences to fill in the blanks.

> *When roses are red*
> *They're ready for plucking;*
> *When a girl turns sixteen*
> *She's ready for . . . 'ere!*

He'd then say, 'I know exactly what you are saying to yourself, you're wrong, I know what you're saying. You wicked lot. You're the sort of people that get me a bad name!'

Max eventually got into trouble with his live BBC radio broadcasts. He was taken off air in 1944 during an unscripted gag about a mountain pass, a girl and a blocked passage and banned for five years.

Cheeky Chappie Max Miller, dominated the music halls of the 1930s–1950s, taking innuendo to new heights

In 1949, the BBC produced 'The Little Green Book', a guide for comedy writers, performers and producers about what was off limits. Under the heading 'Vulgarity', it announced: 'Programmes must at all cost be kept free of crudities, coarseness and innuendo . . . There is an absolute ban on the following: jokes about lavatories, effeminacy in men [and] immorality of any kind.' Also forbidden were 'suggestive references to honeymoon couples, chambermaids, fig leaves, prostitution, ladies' underwear (e.g. winter draws on), animal habits (e.g. rabbits), lodgers and commercial travellers'. 'Extreme care' should be taken with jokes about 'pre-natal influences (e.g. his mother was frightened by a donkey)'. Expletives such as 'God, Good

God, My God, Blast, Hell, Damn, Bloody, Gorblimey and Ruddy' were to be deleted from scripts and 'innocuous expressions substituted'. Chinese laundry jokes 'may be offensive' and jokes like 'enough to make a Maltese Cross' were of 'doubtful value'. Derogatory references to 'Negroes as Niggers' was not allowed but 'Nigger Minstrels is allowed'.

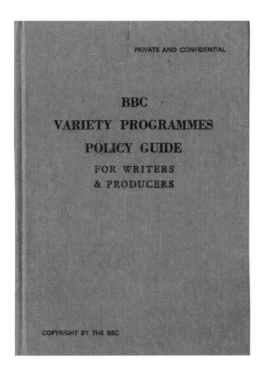

The BBC's 'Little Green Book'

The BBC had arguably lost the battle to protect its listeners even before 'The Little Green Book' was issued. The bawdy banter of the barracks had influenced a generation of servicemen and a new wave of comedians – the likes of Peter Sellers, Spike Milligan and Kenneth Williams – began to write and perform on radio. *The Goon Show*'s Spike Milligan remarked that much of the show's innuendo came from servicemen's jokes, which were understood by most of the cast who had all served as enlisted soldiers and many of the audience, but not by the BBC managers, who were mostly 'officer class'.

The most infamous hotbed of double entendre was the 1960s Radio 2 series *Round the Horne*, in which Kenneth Williams and fellow performers served up a half hour of wordplay, sexual innuendo and nonsense verse every Sunday afternoon. One of the regular characters was an English folk singer, Rambling Syd Rumpo, played by Williams, whose suggestive songs were filled with nonsense words like 'grossets' and 'grommets' and 'nadgers' and 'moolies'.

> *In Hackney Wick there lives a lass*
> *whose grummets would I woggle*
> *Her ganderparts none can surpass*
> *her possett makes me boggle!'*

Music-hall entertainment died out with the success of television, but from the *Carry On* films to TV's *Morecambe and Wise* and *The Two Ronnies*, the format of double entendre family entertainment continued through to the 1970s. The following extract from an episode of *Are You Being Served?*, the BBC department store sitcom, wouldn't have felt out of place in a nineteenth-century musical hall.

Mrs Slocombe: Before we go any further, Mr Rumbold, Miss Brahms and I would like to complain about the state of our drawers. They're a positive disgrace.

Mr Rumbold: Your what, Mrs Slocombe?

Mrs Slocombe: Our drawers. They're sticking. And it's always the same in damp weather.

Mr Rumbold: Really.

Mrs Slocombe: Miss Brahms could hardly shift hers at all just now.

Mr Lucas: No wonder she was late.

Mrs Slocombe: They sent a man who put beeswax on them, but that made them worse.

Mr Rumbold: I'm not surprised.

Miss Brahms: I think they need sandpapering.

There was a defining moment in TV comedy in the early 1980s when the *Not the Nine O'Clock News* team satirized a *Two Ronnies* sketch. Ronnie Barker and Ronnie Corbett's trademark were their innuendo-laden songs.

> *The twittering of the birds all day, the bumblebees at*
> *play.*
> *The twit! The twit! The twit! The twit! The twittering of*
> *the birds all day;*
> *The bum! The bum! The bum! The bum! The bumble-*
> *bees at play . . .*

The *Not the Nine O'Clock News* sketch 'The Two Ninnies' sent up the whole double entendre genre by singing the intended word – or worse. It was knowingly clever in a very 80s way – what would now be called postmodern. It was basically calling the bluff of double entendre.

> *I spend all day just crawling through the grass*
> *Thistles in me hair and bracken up me anus*
> *I'm thrilled to bits to see a pair of tits*
> *And I love to watch the sun go down*
> *Oh vagina oh vagina over Chinatown*

Nowadays, explicit sexual humour raises barely an eyebrow. As the comic writer and performer David Baddiel observes, 'When, twenty years ago, Molly Sugden from *Are You Being Served?* would come on TV and say that it had been raining in the garden and her pussy was soaking wet, it was taken to mean "cat", with a slight overtone of "vagina". Today, it would mean "vagina", with just a tiny undercurrent of "cat".' For the most part, the double entendre has been consigned to early-evening TV sitcoms and that great bastion of bawdy jokes, the pantomime. Stand-up, observational humour is the comedy of choice these days.

And yet . . . living as we do in a society which now talks so openly and frankly about sex, it is odd that we still rather hanker after our double entendres. The BBC Radio 4 programme *I'm Sorry I Haven't a Clue* has been on air since 1972. It's delightfully daft, full of wordplay and puns and absolutely heaving with innuendo – most of it aimed at Samantha, the show's fictional scorer. Of a builder, 'She was pleased to see his tender won but was startled when it suddenly grew to twice its size.'

From office humour to best men's speeches, it seems we still like a bit of innuendo. Long may it continue.

Politeness

Innuendo is one way cultures regulate social behaviour. The language of politeness is another way.

There's an etiquette in grammar which anyone who's learned a foreign language will probably have come across. Most of the Indo-European languages have two levels of grammatical politeness when addressing people. French has the less formal *tu* for children and friends and the formal *vous* for anyone else; Spanish has *tu* and *Usted*, German *du* and *Sie*, Russian ты and Вы. English used to have two levels – *you* and the formal *thou*. *Thou*, *thee* and *thine* have virtually disappeared from day-to-day speech, although they can still be heard in some northern dialects and in a religious setting.

The Japanese language is famous for an extensive grammatical system which is used to express politeness and formality, depending on age, job and experience. There are three levels of politeness, all expressed through different verb endings and alternative expressions. These levels are colloquial, polite and honorific or *keigo* (literally *respectful language*). There are two types of *keigo*: the polite honorific used when addressing someone of higher social status, like the boss at work, teachers or elders, and the humble one when you refer to yourself or family. Forms of *keigo* are found in Korean, Chinese and other Asian languages.

The three golden words of politeness in the English language are *please*, *thank you* and *sorry* and the English – the stereotypical English at least – are the world champions when it comes to using them. We are masters in the arts of supplication, gratitude and apology. 'Sorry to bother you . . . Sorry, but can you help me please? Terribly sorry . . . Thank you. Thank you. Thank you.'

The 1920s guide to proper and correct conduct

Generations of children have been told to 'mind your manners' or 'mind your Ps and Qs'. We presume the 'Ps and Qs' stand for 'pleases and thank yous', although I rather like the theory that the phrase comes from the days of the early printing presses, where mistakes were often made with the typesetting of the lower-case *p*s and *q*s. Anyway, the English are renowned for minding them. We're also rather keen on the matter of table manners, queuing and the art of polite small talk. Discussing the weather is a speciality, as is the avoidance of the contentious subjects of politics, sex and religion.

Other nations have politeness encoded deep into their language and traditions. Iranian/Persian culture has a rather mystifying code of etiquette called *taarof*, which basically means to pay respect to someone. It involves elaborate compliments and praises and requires that you treat your guests and friends better than your own family. *Taarof* is a verbal dance between the person who is offering and the one who is receiving, a volley of insistence and refusal until one of them agrees. *Taarof* governs all levels of daily life, formal and informal – in the market place, shops, restaurants, offices and when entertaining guests at home.

Comedian Omid Djalili was born in London to Iranian parents. Over coffee in his favourite Persian restaurant, he explains how *taarof* works.

'I know, for example, that the lovely Farad here, who owns this restaurant, if I was to come in here, there'd be a little dance where he would give me the best food, and I will say, "Thank you so much, can I have the bill?" And he will say, "My food's not really worth you paying anything." "Please," I say. "No, please, I must pay." He goes, "No, no, no, no, you are a huge person in our community." And I say, "I will, I must pay," and he'll go, "But really, you mustn't pay." And I say, "I really, really must pay." And he'll go, "I'll get the bill then." So he has every intention of charging me, I have every intention of paying, and yet we have this wonderful dance of giving eulogies.'

The same code of politeness operates at home between the host and a guest. Let's imagine your grandmother invites you to her house for dinner. You clear your plate, and she offers you seconds. You're still hungry and you'd love another plateful but you refuse. You've just *taarofed*. The ritual continues, and your grandma offers a second time – you refuse – and then a third time. At this point you accept. It's a laborious, often frustrating, ritual for, of course, you may not have wanted another plateful. You were actually being honest rather than *taarofing*. Some people get round this by asking the guest not to *taarof* ('*taarof nakonid*').

Taarof only works when both sides understand the rules; problems occur when one side doesn't play the game. Omid recalls how his parents were keen to share their *taarofing* culture when they arrived in England from Iran.

'My parents often had English people around, and I'll never forget a very sweet accountant who came to our house around four o'clock and was clearly not hungry. My parents said, "Please, you must have food." And he said, "Oh, I'm not hungry at all." They said, "No, you must eat," and they bring kebabs, they bring rice, and the English person then felt, All right then, I'll have a little bit, just not to appear racist. So he had a little bit to eat, and then my parents said, "Well, take some home," and he did. And as soon as he left, they said, "What a greedy bastard. He took everything. Can you believe it? Doesn't he eat?"'

In *taroof* – unlike some codes of courtesy in countries with caste systems or hierarchical class systems – everyone uses the same ceremony and language with each other, whether they happen to be a prince from the Peacock Throne or a van driver. The aim is to seem as humble as possible, and the language used to achieve this is wonderfully lavish – as Omid demonstrates when the restaurant owner refills his coffee cup.

'I just said thank you very much to Farad,' Omid explains. '*Ghorbanet beram* – which means "May my life be a sacrifice to you." I don't even know him, but I want to sacrifice my life for him! But it's also about humility: like we say *Ghadamet ro Cheshm* – literally, "May you walk on my eyeballs." It means I'm bowing, I want to get as low as I possibly can so you can walk over my brow. It's a way of giving a compliment, you see.'

This emotional, excessive language is not exactly British. E. M. Forster wrote an essay on the character of the English in which he described the reaction of an Indian friend at the end of a week's holiday together. The friend was thoroughly miserable – all happiness, he despaired, had ended. Forster reminded him that they'd be meeting up in a month or two and told him to 'Buck up'. When the two met again, Forster accused his friend of having reacted inappropriately. To this his friend cried, 'What? Do you measure out emotions as if they were potatoes?' Forster explained he was worried that if he poured out his emotions on small occasions, he would have nothing left for the big ones. Emotion, replied his friend, has nothing to do with appropriateness. He was being sincere, he felt deeply and he showed it.

A waiter is clearing away the coffee cups now, and he and Omid are engaging in another round of compliments and counter-compliments. It seems that *taarof* is basically about people feeling good about themselves and at ease in the company of a stranger. It's a way of eroding all the sharp edges and the difficulties of age and gender and education and wealth differences that are bound to arise in any society. But Omid points out that there's a downside to this full-on politeness.

'My wife, who's British, said, "You know, it's very nice, you use such lavish language, you're always complimenting people, but I don't know what's true. You compliment too many people." And you know Iranian culture can be very blunt at the same time; Iranians can't stop themselves. I know if my aunty hasn't seen me for a while, she goes, "Oh my God, you're fat, you've put on so much weight, look at that, oh my God, what are you doing?"'

Perhaps the bluntness is part of the same ritual of establishing an equal footing. Omid isn't convinced. 'To be called a bald, fat fart to your face. That's a bit difficult to take . . .'

Jargon

Anyone who has worked on a filmset will be familiar with the technical language surrounding the actual nuts and bolts of filming which, to an outsider, sounds like coded nonsense. *Sparks, best boys, gaffers, greensmen, grips* and *dolly grips* all have a specialized job to do; their tools are *apple boxes, arcs, bazookas, whips* and *swan necks*. Gobbledegook to most people, but all these words actually allow for a short-hand communication and a very precise use of language. Nearly all professional groups involved in specialized activities, whether they are doctors, lawyers, soldiers or sailors, have their own jargon or terminology.

From film crew to boat crew (it's interesting how film and TV have kept the naval terminology), ocean-racing yachtsman Matt Allen gives a crash course in sailing jargon and its quite bamboozling lexicon for ropes. For starters, a rope on board a yacht is never called simply a rope. Once it's got a designated purpose it's a *line* or, if it's very thick, a *cable*. Lines which are attached to sails to control their shapes are called *sheets*. Stationary lines which support masts are called *standing rigging*; individually they're *shrouds* or *stays*.

Movable lines that control the sails are called *running rigging*. Lines that raise sails are called *halyards*, and those which bring them down are *downhauls*. The bit of rope used to hold the boom down is a *kicker* or a *fall guy*, and a *topping lift* holds the boom up. Clearly, shouting to someone to 'Grab that rope!' could be disastrous.

Matt Allen explains: 'You couldn't operate on a yacht – cruising or racing – without that sort of terminology. Your crew need to know exactly what you mean, which piece of rope to pull or to let off, especially in the heat of the moment. That way they know exactly what you're talking about and there's no confusion.'

We landlubbers try to simplify sailing terminology, to talk about the left side or right side of a boat instead of *port* and *starboard*, but it doesn't work. What we mean by left and right is a position in relation to our bodies (most of us think about which hand we write with). Port and starboard are in relation to the boat itself, so port is the left side of the boat facing forwards and starboard the right side. In fact the word *port* is a good example of how jargon itself changes if it's not clear enough. Sailors used to talk of *starboard* and *larboard*. There must have been frequent misheard shouts in a strong wind, because sailors replaced *larboard* with *port*, as they moored ships on that side at ports.

Nautical terms have permeated our everyday language. Some of the expressions are to do with the technical side of sailing – *ship shape, know the ropes, keel over, be on an even keel, sail close to the wind, take the wind out of someone's sails, make heavy weather, try a different tack, give someone leeway, make headway, give someone a wide berth, trail in someone's wake.*

Other naval phrases are hidden and need some unpicking. To feel *groggy* comes from *grog*, the sailor's daily ration of watered-down rum. If you've drunk too much and you're *three sheets to the wind*, then you're in the same condition as a ship whose lines holding the sails in place are loose – you shudder and roll. *Down the hatch* is another drinking term which comes from the cargo being lowered through the hatch into the ship's hold. *Pipe down*, meaning to be quiet, was the signal at the end of the day for lights out on board. The phrase *no room to swing a cat* is thought to refer to naval floggings using a cat (cat o'nine tails) in the cramped spaces of the old sail ships. Someone in low spirits, who's *in the doldrums*, is experiencing what sailors called the windless, becalming area near the equator. And when the barefooted sailors were called on deck for inspection, they had to line up in neat rows along the seams of the wooden planks and *toe the line*.

The problem with jargon is when it leaves the confines of a particular profession or expertise and is used to communicate with the wider world. That's when the definition of jargon changes.

Chaucer used the word to mean the twittering or chattering of birds, and that's exactly how specialized language of the expert sounds to the punter – a meaningless twitter.

Medical and Legal Jargon

Doctors need to be able to communicate quickly and effectively with other medical professionals. But unlike sailors on a boat or a sparks in a film crew, they have to talk to people outside their specialized group – their patients. It may be a bilateral probital haematoma to them; to us, it's a black eye. Myocardial infarction? That's a heart attack.

Most medical terminology derives from Latin or Greek, so unless you've studied the Classics, jargon 'clues' like *cardio* for heart, *haema* for blood, *tachy* for fast, *hypo* for low and *hyper* for high are meaningless.

In our more patient-centred world, doctors have got much better at dropping the medical jargon and speaking to patients in a language they can understand. They're learning to be more bilingual, shifting from *seborrhoeic dermatitis* to *a touch of dandruff*.

Many of the complicated medical terms have been replaced by another form of jargon – the medical acronym. It's clearer and quicker to use abbreviations for long-worded conditions. DVT is much simpler than deep vein thrombosis or URI for upper respiratory infection; sometimes an abbreviation is used to avoid upset, as in DNR for 'do not resuscitate'. The acronym is also used as a secret code between doctors to talk candidly about the patient, although legislation allowing patients access to their own medical notes may curtail the practice. These acronyms, however, show no evidence of a cramping of style.

FLK: funny-looking kid
ABITHAD: another blithering idiot – thinks he's a doctor
ART: assuming room temperature (recently deceased)
FFFF: female, fat, forty and flatulent
GOK: God only knows
GLM: good-looking mum
TTGA: told to go away
MFC: measure for coffin
NFWP: not for *War and Peace* (dying so no point in starting a long book)

Lawyers too are renowned for baffling their clients with *a prioris*, *ibids*, *idems* and obscure case references. The critic A. P. Rossiter wrote in *Our Living Language* in 1953: 'It strikes everyone as an extreme case of the evils of jargon when a man is tried by a law he can't read, in a court which uses a language he can't understand.'

In English civil courts, attempts have been made to demystify the language with the abolition of some of the most archaic legal jargon and Latin maxims. Since 1999, people bringing cases to court are *claimants*, not *plaintiffs*, a *writ* is called a *claim form*, and *minors* are *children*. The new legal terms are designed to help people understand the law, so *in camera* has become *private*, a *subpoena* is a *witness summons*, and an *Anton Piller*, named after the plaintiff (sorry, claimant) in a court case in the 1970s, is now simply a search and seizure order.

Legalese, the term for legal writing that's designed to be difficult for the layman to read or understand, may be harder to eradicate. The long-winded sentences, countless modifying clauses and complex and often unnecessary vocabulary have been annoying non-lawyers for centuries. Four hundred years ago, Miguel de Cervantes' *Don Quixote* had this to say about legalese: 'But do not give it to a lawyer's clerk to write, for they use a legal hand that Satan himself will not understand.'

" QUID PRO QUO? IT MEANS THEY FUCK US,
WE FUCK THEM. "

*Lawyers are
renowned for
baffling their
clients*

Lawyers want to write contracts which are legally binding and cover all possible contingencies but even they often have trouble decoding the legalese. Documents are peppered with *subsequent to* and *forthwith* and double-barrelled *keep and maintain* and *goods and chattels*. Campaigners for the use of plain language give this example of legalese:

> *Upon any such default, and at any time thereafter, Secured Party may declare the entire balance of the indebtedness secured hereby, plus any other sums owed hereunder, immediately due and payable without demand or notice, less any refund due, and Secured Party shall have all the remedies of the Uniform Commercial Code.*

The plain-language alternative? 'If I break any of the promises in this document, you can demand that I immediately pay all that I owe.'

Plain English in the Workplace

The battle for plain English in the world of politics and government has been raging longer than you might think. In 1948, a British civil servant, Sir Ernest Gowers, was invited by HM Treasury to produce a manual for government officials on how to avoid over-elaborate writing. *Plain Words, a Guide to the Use of English* was so successful that Gowers followed it with *The ABC of Plain Words* and *The Complete Plain Words*.

Sir Ernest cites the example of a circular sent from a government department to its regional offices which began: 'The physical progressing of building cases should be confined to . . . ' Sir Ernest writes:

> *Nobody could say what meaning this was intended to convey unless he held the key. It is not English, except in the sense that the words are English words. They are a group of symbols used in conventional senses known only to the parties to the convention.*

A member of the department explained to him that the phrase meant going to a building site to see how many bricks had been laid since the last visit. 'It may be said that no harm is done,' continues Sir Ernest,

> *because the instruction is not meant to be read by anyone unfamiliar with the departmental jargon. But using jargon is a dangerous habit; it is easy to forget that the public do not understand it, and to slip into the use of it in explaining things to them. If that is done, those seeking enlightenment will find themselves plunged in even deeper obscurity.*

He included a list of 'overworked' words, which he described as 'good and useful . . . when properly used; my worry is only against the temptation to prefer them over other words which would convey better the meaning you want to express'.

More than half a century later, most of the words contained in the list remain firmly entrenched in government-speak: *utilize, envisage, implement, viable, visualize, rendition* . . . But these overworked words are nothing compared to a deluge of non-words which has gripped the English language. This jargon, according to the American poet David Lehman, 'is the verbal sleight of hand that makes the old hat seem newly fashionable; it gives an air of novelty and specious profundity to ideas that, if stated directly, would seem superficial, stale, frivolous, or false'. It's called, variously, corporate-speak or bureaucratese or offlish (for office English), and its terminology of *blue-sky thinking* and *benchmarking* and *thinking outside the box* and *synergizing* and *conditionality* has spread throughout the world of corporations and government departments and offices. Our language thrives on innovation, but the baffling phrases, power words, tortured verbs and pointless adages of corporate jargon have little to recommend them. This jargon appears to be neither inclusive nor humorous nor very clever.

A wadge of gung-ho transitive verbs are favoured: *to action, to incentivize, to leverage, to strategize, to downsize.* Everything is upbeat. Problems aren't problems, they're *challenges*; commitment is *110 per cent*; anything done in the future is on a *go-forward basis.* Other monstrosities include *I don't have the bandwidth to deal with the situation* rather than 'I don't have the time'; *end-user perspective* instead of 'what the customer thinks'; *cascade down information* for simply sending a memo.

A new office pastime has been created. Employees play Buzzword – or Bullshit – Bingo in the boardroom, ticking off a predetermined list of jargon words uttered during the meeting. The first person to have a full card is supposed to yell 'Bingo!'

Corporate-speak is inveigling itself into every corner of officialdom. The jobs section in a British newspaper had the following advert: 'Proactive, self-starting facilitator required to empower cohorts of students and enable them to access the curriculum.' That's a teacher, to you and me.

There's no British institution that has ridiculed pretentious or obfuscatory language more than *Private Eye*, the satirical magazine. The magazine's editor, Ian Hislop, has tracked and excoriated the rise and rise of business and political jargon since he took the job in 1986.

> *Using language as a way of obscuring the truth rather than revealing the truth is always dangerous, and so I think that's part of the point of attempting always to monitor these excesses. And English is a very precise language. It can be used to convey anything beautifully but it's also very amenable to nonsense . . . The British are obsessed with their own language, and* Private Eye *gives them a way of monitoring it. So a lot of these columns were actually started by readers just saying, 'Have you noticed that everybody is using the word "solutions"?' You can't get your windows replaced now, someone does 'window solutions'. You can't get a garden hose, you have 'water irrigation domestic solutions'.*

Hislop has a theory on how management-speak has spread like a virus through institutions like the BBC, the NHS, the civil service and local government.

> *It starts in management consultancies, which are firms designed to make a science out of what used to be an art or common sense – management, dealing with people. Management consultants make this into a science. You hire management consultants usually for two reasons. One is you want to sack people and you don't do it yourself or, two, you want to create verbiage to describe non-existent jobs. So you're either getting rid of people who do a real job or you're inventing non-jobs. And the jargon does perfectly for both of those. So the people in non-jobs can send each other memos about rolling out milestones and delivery and competence, and the people who are being sacked are told that they've been restructured.*

The language of management seems to have deteriorated at such a breakneck speed in terms of its warmth and emotional directness that it's hard to imagine it getting any more impersonal. Ian Hislop says companies are aware of the problems.

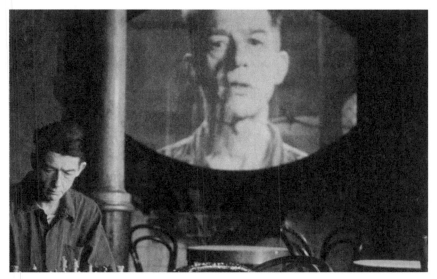

John Hurt as Winston Smith in the film version of 1984

What amuses me is the same management who basically are bringing in systems to make sure people fall apart then are told there's no bonding going on. So they have to organize paintballing weekends and start bringing in members of the SAS to give talks about getting across bridges without using rope and forcing people to socialize because they've become so disparate inside the office.

In the brave new world of management speak, harsh realities get hidden. This is the language of doublespeak, deliberately euphemistic, ambiguous or obscure. In some cases it actually reverses the meanings of words. 'Doublespeak' is a term which can be traced back to George Orwell, who invented the words *doublethink* and *newspeak* for his novel *1984*.

American linguist William Lutz writes:

> *Doublespeak is language which pretends to communicate but doesn't. It is language which makes the bad seem good, the negative seem positive, the unpleasant seem unattractive, or at least tolerable. It is language which avoids, shifts or denies responsibility; language which is at variance with its real or purported meaning. It is language which conceals or prevents thought. ('Doubts about Doublespeak', State Government News, 1993)*

Doublespeak with Political Intent

The uglier side of doublespeak is its use as camouflage to hide the reality. In the world of business, euphemisms are bald: workers aren't sacked, they're *down-* or *right-sized*, *derecruited* or *involuntarily terminated*. Companies don't suffer losses, they have *negative cash flows* or *downward adjustments* or *negative growths*.

Most disturbing of all is how the doublespeak of business and management has been adopted by governments and politicians who use deliberately ambiguous phrases to make us feel better about politically sensitive subjects like war or killings or torture. To kill becomes to *take down*, *take out* or *neutralize*, or the 'unlawful and arbitrary deprivation of life' (US State Department annual report 1984). Civilian casualties are *collateral damage*; an escalation in fighting is a *surge*; state kidnapping for the purposes of torture becomes *rendition*; a terrorist furthering state interests is a *freedom fighter*; genocide is changed to *ethnic cleansing*. Weapons are *assets*; nuclear weapons are *nuclear deterrents*. Torture is *enhanced coercion interrogation technique*.

Political doublespeak isn't a new phenomenon. It's most infamous use is the terminology of the Nazis to describe the systematic extermination of Jews – *the Final Solution*. Hitler used euphemisms to dehumanize and make the unacceptable acceptable. He spoke about the need to *purify* and *cleanse*, to rid the Reich of the Jewish *vermin* and to *decontaminate* or *disinfect* the Reich of the Jewish *bacillus*. Instead of kill or murder, expressions like *special treatment, evacuation, resettlement* or *conveyed to special measure* were used. The planned killing of handicapped people was *euthanasia* or *mercy death*. Poland, with its death camps, was called the *Jewish resettlement region*; gas chambers were *bathhouses*, and mobile gas chambers were *auxiliary equipment* or *delousing vans*.

Victor Klamperer, a German Jewish Professor of Literature, documented in his book *LTI, Lingua Tertii Imperii* (The Language of the Third Reich) the daily mental corruption of the German people through language. Klamperer escaped the gas chambers because his wife was Aryan.

> *Nazism permeated the flesh and blood of the people through single words, idioms and sentence structures which were imposed on them in a million repetitions and taken on board mechanically and unconsciously ... Language does not simply write and think for me, it also increasingly dictates my feelings and governs my entire spiritual being the more unquestioningly and unconsciously I abandon myself to it. And what happens if the cultivated language is made up of poisonous elements or has been made the bearer of poisons? Words can be like tiny doses of arsenic: they are swallowed unnoticed, appear to have no effect, and then after a little time the toxic reaction sets in after all.*

The city of Leipzig lies in what we used to call East Germany. This part of Germany suffered two extreme regimes in the last century – the fascist Nazis and then the communists, who controlled what was then called the GDR (German Democratic Republic) from the end of the Second World War until 1989 and the fall of the Berlin Wall. Like the Nazis, the communists were expert at using language to control and subdue, changing words and changing the meaning of words to suit their political purposes. In the former headquarters of the notorious Staatssicherheit – the Stasi, probably the most famous secret police after the KGB – political satirist Gunter Böhnke recalls that party officials called the secret police 'die Sicherheit' – the security. Ordinary people nicknamed it 'die Stasi', making it sound feminine – a bit like us calling the police 'the She Police', he says. It was the public's small way of showing defiance, making the Stasi seem less frightening.

The Berlin Wall was officially called the 'Anti-Fascist Barrier' by the GDR authorities, but, as Gunter recalls, 'the barbed wire was not facing the West but to the East. Everybody could see that the enemies could come in. But you were not allowed to go out.'

The GDR was nothing like the Nazi regime in terms of terror and murder. This was a much more insidious tyranny in which all conversations were monitored by an army of citizen spies – some estimates say as many as one for every six and a half members of the population. Gunter describes it as a sort of mind control where even telling a joke was dangerous.

'My mother lost her purse with a number of Ulbricht [Walter Ulbricht, GDR leader] jokes in it, and for months on end with every ringing of the bell, we thought the Stasi will come and take our mother because of these jokes. It was enough just to tell a joke. There was a teacher of Russian who told a joke to his colleagues about Krushchev. Somebody reported him to the Stasi, and he was sacked from school and had to work in a chemical factory. Very hard, very dirty work. This was in 1962.'

Gunter himself was allowed to perform comedy cabaret shows as a 'steam valve' as long as the jokes didn't attack senior party officials. George Orwell wrote: 'Every joke is a tiny revolution,' and despite the danger from informants, political jokes thrived. Some of the more critical ones were known as 'five-year jokes' – three years in prison for the person telling it and two years for everyone else who listened and laughed.

In the 1970s and 80s the GDR leader, Erich Honecker, was the target of a number of jokes. This one features in the 2006 Oscar-winning film *The Lives of Others*:

Erich Honecker arrives at his office early one morning. Opening his window he sees the sun and says, 'Good morning, dear sun.'

The sun replies, 'Good morning, dear Erich!'

Honecker gets on with his work and at noon he opens the window and says 'Good day, dear sun.'

And the sun replies, 'Good day, dear Erich.'

In the evening, as he heads out of his office, Erich goes again to the window and says, 'Good evening, dear sun.' The sun is silent. Honecker says again, 'Good evening, dear sun! What's the matter?'

The sun replies, 'Kiss my arse! I'm in the West now.'

Modern Taboos

Given the atrocities in our history perpetrated by one group against another, it's not surprising that many governments and institutions have tried to legislate against so-called hate speech – disparaging remarks about religions, ethnicities, nationalities, sexualities and genders. But who defines hate speech? Can we prevent hate speech without encroaching on freedom of speech? And does making something taboo merely give it more power?

This chapter began with the Seven Words You Cannot Say – swear words which our forefathers would physically recoil at but whose power to offend this generation is waning. Today, old taboos have been replaced by new ones, which language and humour have to negotiate. These are the taboos of homosexuality and disability and – the issue which probably makes us most uncomfortable and which hardly bothered previous generations – race. When we joke about race we tread on eggshells. There are words we just don't use. And the most offensive one is probably the word *nigger*. F-words and c-words cause only mild ripples these days, but the n-word is extremely loaded.

It wasn't always like this. Older readers may remember merrily reciting a children's counting rhyme:

Eeny, meeny, miney, mo,
Catch a nigger by the toe;
If he squeals let him go,
Eeny, meeny, miney, mo.

It was acceptable right into the 1970s, just as collecting the Golliwog stickers off the back of jars of Robertson's jam and sending them off for a Golliwog badge seemed an innocent hobby. There was little fuss when Agatha Christie published her bestselling detective thriller *Ten Little Niggers* in 1939 – although the Americans brought it out as *And Then There Were None* the following year. British publishers didn't change the title until 1985.

A proposed remake of the 1955 film *The Dam Busters* has stepped into a quagmire of political correctness. One of the film's main characters was Guy Gibson, the RAF commander of the British mission that destroyed German dams with 'bouncing' bombs in the Second World War Two. Gibson had a black Labrador called Nigger – a common enough name for a black dog in those days; it was also the radio codeword used to report the success of Gibson's squadron on one of the targets. ITV broadcast a censored version of the original film in 1999 with all 'Nigger' utterances deleted; the Americans dubbed over 'Nigger' to make it 'Trigger'. So should the film-makers stick to the facts or alter history, changing a name that was perfectly acceptable in the 1940s and 50s so as not to offend people today? As the director of the remake, Peter Jackson, notes, 'We're in a no-win, damned if you do and damned if you don't scenario.'

A much more objectionable rewriting of history in the name of political correctness was the publication of an edition of Mark Twain's *Huckleberry Finn* in 2011 with every mention of 'nigger' excised and replaced – over 200 times – with the word *slave*. It's been described as a kind of ethnic cleansing, a whitewashing of the fact that black people in the American South in the mid nineteenth century were referred to as 'niggers'. And a complete failure to understand that *Huckleberry Finn* is actually anti-racist.

The Americans are acutely sensitive about the n-word; it makes them linguistically twitchy. In 1999, David Howard, a white aide to the black mayor of Washington, DC, was having a financial discussion

*RAF crew and Guy Gibson making a fuss of their black Labrador dog,
Nigger, who is wearing an Iron Cross*

with a black colleague when he talked about being 'niggardly' with
the budget. Now *niggardly* means 'miserly', probably from the Old
Norse *hnøgger* for stingy. But it's an uncommon word and it sounds
like *nigger*, so was interpreted as a racial slur. A complaint was
lodged, and Howard tendered his resignation. It was accepted with
alacrity by the mayor. A national debate on political correctness
ensued, with the chairman of the African-American civil rights group
Julian Bond, opining: 'David Howard should not have quit. Mayor
Williams should bring him back – and order dictionaries issued to all
staff who need them . . . Seems to me the mayor has been niggardly
in his judgment on the issue.' David Howard was brought back to
work in the mayor's office.

Racial linguistic sensibility is clearly not so acute in Britain, where
society was largely white until the 1940s. Mass Afro-Caribbean immigration began after the Second World War, but it took decades before
the word *nigger* was generally accepted as a racial slur. *Love Thy*

Neighbour was a sitcom on ITV in the 1970s about a white family and a black family who lived next door to each other. Eddie, the white male character, talked about 'sambos' and 'nig-nogs'. The script-writers claimed that *Love Thy Neighbour* was an attempt to address some of the issues raised by a growing immigrant population in Britain. The characters may have been racist, they argued, but the show wasn't.

Black and gay stand-up comedian Stephen K. Amos was a school-boy in south London when *Love Thy Neighbour* was broadcast. 'I'd go to school on the Monday and be called a nig-nog because they'd see it on the show . . . I didn't know I was a nig-nog until my class-mates told me I was.'

The interesting thing about today's taboo words is that it's seen as okay to use them if you're part of the particular community. So Jews are allowed to say *kyke*; Stephen Amos is gay and feels comfortable about using the word *queer*. A lot of black comedians like the American Chris Rock use the n-word quite freely, and there's even a group called NWA, Niggers with Attitude. Stephen says he doesn't use the n-word personally but understands why

Love Thy Neighbour, *1972*

others might want to reclaim the word. What's more important for him is that taboo words cause comedians to think before they speak.

'When people say political correctness has gone mad, I really get offended by that term because I don't think it's being politically correct if you have to think before you speak, if you have to think before offending people. If you're a clever comedian and you want to upset the apple cart then, yes, do that, but do it in a way which makes us all think, not just by throwing in a word or having a go at a community or at disabled kids. If there isn't a purpose or a point, there's no point, because we could all do that.'

Slang

Language is in a constant state of change and reinvention, and slang plays a vital role in this evolution. Slang is described in early editions of the *Oxford English Dictionary* as the 'language of a low and vulgar type . . . consisting either of new words or of current words employed in some special sense'. The origin of the word *slang* is unknown. It doesn't show up until the middle of the eighteenth century, when it was used to refer to the 'special vocabulary of tramps or thieves'. Before then it was called *cant* or *vulgar language*.

Street slang, rhyming slang, back slang, teenage slang, text slang – there's an abundance of slang in the English language, and the strong feelings it generates are nothing new. From the sixteenth century onwards, people were railing against use of the vulgar tongue. In 1621, John Milton's headmaster, Alexander Gil, wrote about the cant speech of 'the dirtiest dregs of the wandering beggars'. In his study of grammar and dialects, *Logonomia Anglica*, he described cant as 'that poisonous and most stinking ulcer of our state'.

The satirist Jonathan Swift was a passionate advocate of the need to purify the English language. In an article published in the *Tatler* in 1710 he attacked what he called 'the continual corruption of our

English tongue', not by the common people but by the writers and poets of the age. Swift denounced the use of abbreviations, in which only the first part of a word was used, and also:

> *the choice of certain words invented by some pretty fellows:*
> *such as banter, bamboozle, country put, and kidney, as it*
> *is there applied; some of which are now struggling for the*
> *vogue, and others are in possession of it. I have done my*
> *utmost for some years past to stop the progress of mobb*
> *and banter, but have been plainly borne down by numbers,*
> *and betrayed by those who promised to assist me.*

Swift clearly failed to stop the entry of words like *mob* (from *mobile vulgus*, the Latin for fickle crowd) and *banter* into our everyday language.

The first substantial dictionary of slang, *A Classical Dictionary of the Vulgar Tongue*, was published in 1785 by Francis Grose. He was a former soldier, innkeeper and champion drinker who collected slang from all corners of society – sailors, tradesmen, prostitutes, pickpockets and craftsmen. He and his assistant are said to have walked the slums of London at night, noting down the cant words spoken in the drinking dens and brothels. His dictionary included over 3,000 entries. Some of them are familiar – *hen-pecked, topsy-turvy, brat, sheepish* (for bashful) and *carrots* (for red hair). Others are, rather sadly, obsolete. Words like *circumbendibus* (a wandering path or story) or *scandalbroth* (tea).

What Grose calls vulgar, we would probably call slang. His dictionary meaning for *devilish*, for instance, reads: 'an epithet which in the English vulgar language is made to agree with every quality or thing; as, devilish bad, devilish good; devilish sick, devilish well; devilish sweet, devilish sour; devilish hot, devilish cold, &c. &c.' Many of the words reflected the seamier side of life. A *covent garden nun* was a prostitute, and the delightful-sounding *scotch warming pan* was a word for 'a wench, also a fart'. There are plenty of rude words. *Shag, hump* and *screw* are all there for copulation. Less familiar are *bum fodder* for toilet paper and *double jugg* for a man's bottom. And there's one listed simply as 'c—t: a nasty name for a nasty thing'; elsewhere he refers to it as 'the monosyllable'.

CAPTAIN FRANCIS GROSE
BY HIMSELF

*Francis Grose created the first
dictionary of slang in 1785*

Unlike Swift, Grose believed that the rich abundance of slang words in the English language was something to be celebrated: 'The freedom of thought and speech, arising from, and privileged by, our constitution, gives a force and poignancy to the expressions of our common people, not to be found under arbitrary governments.'

There's a fascinating postscript to this larger-than-life character. Grose met the Scottish poet Robert Burns when he was in Scotland, drawing sketches and collecting material for a book on local antiquities. They got on famously – Burns wrote: 'I have never seen a man of more original observation, anecdote and remark' – and Grose agreed to include a drawing of Alloway Kirk in his forthcoming volume, if Burns would provide a witch tale to accompany it. In 1790 Burns sent him the rhyming tale of 'Tam O'Shanter' – arguably one of the best examples of narrative poetry in the English language.

Back-slang

The 'secret tongue' of the costermongers, the mobile fruit and veg sellers, was back-slang – essentially pronouncing a word backwards. There were tens of thousands of 'costers' in Victorian London, with a reputation, according to a book published in 1859 by John Camden Hotton entitled *A Dictionary of Slang, Cant and Vulgar Words*, of 'low habits, general improvidence and their use of a peculiar slang language'. They were a tight-knit community with a common enemy – the police – and seemed to have developed back-slang, a private language which the punters and non-locals couldn't understand. The back-slang was used mostly for words involved in their trade and everyday life – coins, vegetables, fruit and police. So *dunop* was a pound, *yennep* a penny, *rape* a pear, *storrac* carrots, *spinsrap* parsnips, *slop* a policeman. A costermonger told Henry

Taken 4ª Feb. 1893.

Costermongers loved to use the secret language of back-slang

Mayhew, author of the study *London Labour and the London Poor* (1851): 'I likes a top o' reeb.' Almost all the words have become obsolete, except *yob*, which is, of course, 'boy' backwards.

Back-slang was gradually abandoned by the costers and replaced by rhyming slang. The butchers took it up, and by the twentieth century back-slang was regarded as entirely their secret language.

What is it about butchers that makes them so secretive? Butchers in Paris and Lyon developed their own secret language in the mid nineteenth century as well. It was a much more complicated slang called *loucherbem*, closer to Pig Latin than back-slang. The first consonant of each word is moved to the end, a suffix such as -em is added and the letter L is added to the start of the new word. Thus *boucher* (butcher in French) becomes *loucherbem*.

Cockney Rhyming Slang

Secret languages are the stuff of childhood. Generations of school-boys have thrilled to the mystery of writing notes in invisible ink and speaking in a code which no one outside their inner circle of chums understood. Pig Latin was a language game beloved of school-children. It wasn't really Latin (just sounded a bit like it) and involved putting the first letter of a word at the end and then adding -ay. 'Owhay oday ouyay oday?' (How do you do?'). Pig Greek – or ubbi dubbi – is another one; also aigy paigy, Double Dutch and gibberish.

These secret languages are essentially games, abandoned in adulthood. But there are other ingenious, covert languages which have developed within a group or a community, often to hide illicit practices or allow coded talk about others without them knowing; in some instances, they've spread to become part of the general vocabulary. The argot which we are all most familiar with – especially through TV shows like *Minder*, *Only Fools and Horses* and *EastEnders* – is Cockney rhyming slang.

Del Boy carrying on the Cockney tradition

Much more fun than Pig Latin, rhyming slang is a glorious feast of linguistic gymnastics, peculiar to the English language and prevalent in the East End of London in the second half of the nineteenth century. The construction involves replacing a word (let's say 'feet') with a rhyming phrase ('plates of meat') and then dropping the rhyming part of the phrase ('meat' goes, 'plates' stays). In practical terms it means that 'feet' become 'plates' and unless you're familiar with the rhyme, the original word is hidden.

Have a go at translating this:

I had a Jane down the frog with a septic, his trouble and their dustbin lid. Would you Adam and Eve it? My old china was wearing a syrup under his titfer, a whistle, a Peckham and a pair of churches.

Translation: I had a wander (*Jane Fonda*) down the road (*frog and toad*) with an American (*septic tank* – Yank), his wife

(*trouble and strife*) and their kid (*dustbin lid*). Would you believe
(*Adam and Eve*) it? My old mate (*china plate*) was wearing a
wig (*syrup of fig*) under his hat (*tit for tat*), a suit (*whistle and
flute*), a tie (*Peckham Rye*) and a pair of shoes (*church pews*).

No one knows for sure when and where rhyming slang originated,
but it was certainly flourishing in early Victorian England. Henry
Mayhew noted: 'The new style of cadgers' [street sellers'] cant is all
done on the rhyming principle.' John Camden Hotton informs us
that rhyming slang originated in the 1840s with 'the wandering tribes
of London'. Hotton is adamant that the rhyming wasn't invented
by costermongers – who took up the rhyming slang later – but by
two other types of street traders.

> *There exists in London a singular tribe of men, known
> amongst the 'fraternity of vagabonds' as Chaunters and
> Patterers. Both classes are great talkers. The first sing or
> chaunt through the public thoroughfare ballads – political
> and humorous – carols, dying speeches, and the various
> other kinds of gallows and street literature. The second
> deliver street orations on grease-removing compounds,
> plating powders, high polishing blacking, and the thou-
> sand and more wonderful pennyworths that are retailed
> to gaping mobs from a London kerb stone.*
>
> *They are quite a distinct tribe from the costermongers;
> indeed, amongst tramps, they term themselves the 'harri-
> stocrats of the streets,' and boast that they live by their
> intellects. Like the costermongers, however, they have a
> secret tongue or Cant speech, known only to each other.*
>
> *This cant . . . is known in Seven Dials [a notoriously
> disreputable part of London] as the Rhyming Slang, or
> the substitution of words and sentences which rhyme with
> other words intended to be kept secret.*

Hotton's *Glossary* included rhyming slang still in currency today:
apples and pairs – stairs; *elephant's trunk* – drunk; *pen and ink* –
stink; *mince pies* – eyes; *macaroni* – a pony; *sugar and honey*

– money; other ones, like *Duke of York* – take a walk; and *Top of Rome* – home, have disappeared.

It was called Cockney rhyming slang but really it was a Londoners' slang and especially working-class Londoners. The word Cockney is thought to have derived from *cockeney*, a fourteenth-century word used to describe both a misshapen egg (hence a cock's egg) and a spoilt, 'cockered' child. By the early seventeenth century, the two meanings of being odd and being spoilt appear to have merged into a contemptuous name used by country folk for a soft, puny townsperson, typically a Londoner. Another derivation for the word was suggested by Frances Grose in his *A Classical Dictionary of the Vulgar Tongue*:

> *A citizen of London, being in the country, and hearing a horse neigh, exclaimed, Lord! how that horse laughs! A by-stander telling him that noise was called Neighing, the next morning, when the cock crowed, the citizen to shew he had not forgot what was told him, cried out, Do you hear how the Cock Neighs?*

So a Cockney was a Londoner in general, or more specifically, according to John Minsheu in his *Ductor in linguas* (Guide into Tongues) dictionary in 1617, 'one born within the sound of Bow bell, that is in the City of London'. The Bow Bells were in the church of St Mary-le-Bow in Cheapside, an area that today is largely non-residential. The bombing of the East End of London during the Second World War meant the migration of huge numbers of traditional Cockneys to the new towns on the outskirts of Greater London. You're more likely to hear old-style Cockney rhyming slang in Basildon or Harlow in Essex and in parts of Hertfordshire than in Cheapside or Whitechapel. In turn, waves of immigration into London's East End – most recently Bengali – have provided a melting pot mixture of different languages. East end teenagers today talk about *skets* not *eggs and kippers* (slippers) or *creps* rather than *Gloria Gaynors* (trainers). And life isn't so much *Robin Hood* (good) as *Nang!* Years before the traditional Cockney-speakers were displaced to the London outskirts, Cockney rhyming slang was

Born within the sound of Bow Bells and you are officially a Cockney

spreading itself throughout Britain and beyond. Observers commented on how the language of the first convict settlers to Australia was that of Cockney London, and the exuberance of today's Australian slang has undoubtedly been influenced by it. In Britain, some of the words and phrases have lodged so firmly in our everyday language that we'd be hard put to recognize them as original rhyming slang.

Rabbit on (*rabbit and pork* – talk)

Use your loaf (*loaf of bread* – head)

Have a butchers (*butcher's hook* – look)

Don't say a dicky bird (word)

Blow a raspberry (*raspberry tart* – fart)

Tell a porky (*pork pie* – lie)

On your tod (*Tod Sloan* – alone)

Tod Sloan, the American jockey, introduced the 'monkey crouch'

There are snippets of social history hidden in some of these phrases. Tod Sloan was an American jockey who became an international celebrity at the turn of the twentieth century. He introduced the 'monkey crouch' forward seat riding position to horse racing and rode British winners at Newmarket and Ascot. The Broadway hit song 'Yankee Doodle Boy' was about him: 'Yankee Doodle came to London, just to ride the ponies . . .'

Sometimes with slang, the more people that understand it, the more you have to change it. If they work out that *china plate* means 'mate', you drop the rhyme word and just say *china*. Or if they unravel what to kick someone in the *cobblers awls* means, then you shorten it to *cobblers*. Or the *orchestras* (stalls). Or the *Niagras* (Falls). The rhyming slang for 'arse' is especially confusing. The first rhyme was *bottle and glass*, then simply *bottle*, then *Aristotle* and finally *aris*. This is occasionally extended further to *April* (in Paris) and you're left with a sentence like 'She fell on her *April*.'

Rhyming slang words can be passed down through the generations or they might be hugely popular for a few years and then be overtaken by the next fad. Some words depended on knowledge of

London. If someone threatened to kick you in the *Hampsteads*, you had to be familiar with Hampstead Heath to know it meant teeth. People still talk about going to the barbers to get their *barnet* cut. This word for hair has been around since at least the 1850s and comes from the popular Barnet Fair in north London. *Hampton Wick* (near Teddington) is rhyming slang for prick or dick and is often just *Hampton* or *wick*. *He gets on my wick* – meaning he's annoying – is one of those innocuous phrases that's actually rather rude. Spike Milligan managed to introduce a character called Captain Hugh Jampton into a *Goon Show* episode in 1958. The BBC banned a further appearance when they worked it out. The *Carry On* films had a field day. *Lord Hampton of Wick* appeared in *Carry on Henry*, and the nurse and doctor *Carry On* films were based at Long Hampton Hospital. And the first of the serials within *The Two Ronnies* TV show was called 'Hampton Wick'.

By the mid twentieth century many rhyming slang expressions used the names of contemporary personalities, especially actors and performers. Ruby Murray was a Belfast-born singer, popular in the 1950s at the same time as Indian restaurants were becoming widespread. So today her name *Ruby* lives on as rhyming slang for a curry. 'I'm going for a *Ruby*.' One of the rhyming slangs for deaf is *Mutt and Jeff* or simply *mutton*. I wonder how many people using the phrase know it comes from characters in an American comic strip created by Bud Fisher in 1907. Your grandmother might have got new *Teds* or *Ted Heaths* (false teeth). Or, referring to a more recent personality, you could say she's got a nice pair of *Penelopes* (Keith). A pair of knickers are *Alan Whickers* or simply *Alans*, as in the film *Lock, Stock and Two Smoking Barrels*: 'All right, all right, keep your Alans on!'

Rhyming slang can evolve more than one meaning. Woe betide if you have a name that rhymes easily. Actor Gregory Peck's name was both 'neck' – get that down your *Gregory* (or as characters in the *Minder* TV show were constantly urging, 'Let's get a Ruby down your Gregory'), and 'cheque' – I'm going to cash a *Gregory*. It can also be used in the plural, as in wearing my *gregs* – 'specs'. A *Melvyn*, from the arts broadcaster Melvyn Bragg, has been used at various times to mean 'shag', 'fag' and 'slag'. A *Melvyn* is not to be confused with a *melvin*, which, according to the *Historical Dictionary of American*

Slang, means 'pulling someone's pants up sharply to wedge them between the buttocks'.

Television and films, the internet, texting and tweeting, together with a much greater racial and cultural mixing pot, ensure that most of the new rhyming slangs are ephemeral, discarded as quickly as the next celebrity or fad comes around. And yet, to mix a metaphor, amidst the chaff there are some gems. A market stall holder in London's Soho was heard to say, 'Oh, it's the tourists . . . I'm not Listerine but they get on my goat.' Rhyming slang for American is *septic* (from septic tank – Yank). So if you're *Listerine* (a mouth-wash), you're *anti-septic*, i.e. anti-American.

Some Cockney cabbies talk about how the Cockney they grew up speaking is gradually fading away.

'Well, we're losing it, aren't we? But then, our way of life is changing, isn't it? See, we used to have stall holders, in the markets, and their children would come up and they would learn the patter, and that was handed down. Well, now them children are going to university or whatever because they always try and do better for their children. You've only gotta look at your kids. Your kids are picking up the hip-hop type of language. As much as you try, when they're at school they're picking up the various patters. In the same way we used to, 'cos it was always Cockney that was spoken.'

One cabbie says his children still recognize some Cockney rhyming words but, as all slang does, they've evolved. So instead of asking his daughter whether she's having a *tin*, meaning laugh (tin bath), he'll ask her whether she's having a *bubble*. The old tin bath isn't around any more.

These cabbies reckon their rhyming acts as a sort of safety valve, helping to take the edge off offensive or racist labelling. They describe their fares as *seppos* (Yanks) or *tiddly winks* (Chinks), or *fourbytwos*, *tinlids* or *front wheel skids* (Jews and Yids).

'A Cockney has got a cheerful way about him. So when he's saying it, it's not in an offensive manner, it's in good fun, it's always with a good humour. It's always with a smile on his face.'

Rhyming slang is a bit like one of those minced oaths. We know we can't say the taboo word so we come up with a cheerful alternative which makes us smile. Most of the time.

Polari

In rhyming slang a gay man might be a *ginger* (beer – queer) or *King Lear* (queer again) or *iron* (hoof – poof). But gay men know all about secret codes themselves. Back in the 1960s, homosexuality was still a crime, and gay men in London used a secret language to identity and secretly communicate with each other. The language was called Polari and it had a vocabulary of about twenty key words, mainly describing people's looks, clothes and sexual availability: *bona* for good; *omi* man; *palone* woman; *omi-palone* gay man; *eek* face (from back-slang *ecaf*); *fabulosa* wonderful; *riah* hair (back-slang); *vada* to look; *naff* not available for fucking; *camp* effeminate (also from Kamp, acronym for 'known as male prostitute'); *zhoosh* to fix or tidy.

Gay journalist Peter Burton included an example of Polari in his autoboigraphy *Parallel Lives*.

> As feely ommes . . . we would zhoosh our riah, powder our eeks, climb into our bona new drag, don our batts and troll off to some bona bijou bar. In the bar we would stand around with our sisters, vada the bona cartes on the butch omme ajax who, if we fluttered our ogle riahs at him sweetly, might just troll over to offer a light for the unlit vogue clenched between our teeth.
>
> (As young men . . . we would style our hair, powder our faces, climb into our fabulous new clothes, don our shoes and wander/walk off to some fabulous little bar. In the bar we would stand around with our gay companions, look at the fabulous genitals on the butch man near by who, if we fluttered our eyelashes at him sweetly, might just wander/walk over to offer a light for the unlit cigarette clenched between our teeth.)

The origins of Polari are unclear. It's a linguistic mongrel, borrowing words from Occitan, Romany, Shelta (the cant of the Irish tinkers), Yiddish, back-slang and rhyming slang – all interspersed with words of Italian origin. One theory is that Polari – from the

Italian *parlare*, to talk – was the lingua franca of seafarers and traders around the Mediterranean ports in the Middle Ages. It made its way to Britain via travelling circuses and fairgrounds and was used widely on board British Merchant Navy ships.

Between the 1930s and 1970s, Polari was used in theatres and private gay drinking clubs, especially in London. In 1965, Polari came to the attention of a much wider audience with the arrival of a new BBC radio comedy programme, *Round the Horne*. More than 9 million people tuned in every Sunday afternoon to listen to the sketches, one of which featured two camp out-of-work actors called Julian and Sandy, played by Hugh Paddick and Kenneth Williams. Their torrent of double entendre and innuendo included Polari terms which would have sounded like gibberish to most people but which regular listeners learned to decipher. The host, Kenneth Horne, would visit a new enterprise each week – Bona Pets or Bona Ballet or Bona Books or something – and enter saying, 'Hello, is there anybody here?' He'd be greeted by the camp duo. 'Hello, I'm Julian, and this is my friend Sandy.' 'Oh, Mr Horne, how *bona* [good] to *vada* [see] your *dolly* [pretty] old *eek* [face].'

In the sketch 'Bona Law', Barry Took and Marty Feldman included the line: 'Omes and palones of the jury, vada well at the eek of the poor ome who stands before you, his lallies trembling' ('Men and women of the jury, look well at the face of the poor man who stands before you, his legs trembling'). Barry Took explained later that he had learned some of the Polari words during his time as a music-hall comic in the West End. He said Kenneth Williams and Hugh Paddick were always speaking Polari to one another and would sometimes adlib the sketches to include more Polari.

Some of the material was really quite risqué. In one episode Sandy refers to Julian's skill at the piano as 'a miracle of dexterity at the cottage upright'. Only the Polari cognoscenti were likely to have known that a *cottage* was the term for a public toilet where men met for sex and *upright* meant an erection.

By the end of the 1960s, Polari was in decline – partly as a result of the decriminalization of homosexuality and the advent of gay liberation and in part, no doubt, to the success of *Round the Horne*. The secret language was no longer secret. Indeed, some Polari and words have entered into mainstream language: *butch*, *naff*, *queen*, *mince*, *camp*, *drag*, *fab*, *dishy*, *butch*, *bijou*, *savvy*, *scarper*, *tat* and *bevvy*.

Polari may not be used as a secret language by the gay community any more but it is seen as playing an important part in gay cultural history. The Sisters of Perpetual Indulgence, an order of gay and lesbian nuns and monks, translated the King James Bible into Polari in 2003 and posted it on the internet.

> *In the beginning Gloria created the heaven and the earth.*
>
> *And the earth was nanti form, and void; and munge was upon the eke of the deep. And the fairy of Gloria trolled upon the eke of the aquas.*
>
> *And Gloria cackled, Let there be sparkie: and there was sparkle*
>
> *And Gloria vardad the sparkle, that it was bona: and Gloria medzered the sparkle from the munge.*

Round the Horne, *full of risqué Polari and double entendre*

Aussie Slang

It's impossible to write about the uses and abuses of language without mentioning Australia – for it is here we find some of the most playful, colourful, sometimes vulgar uses of the English language. If Shakespeare walked this earth today, he'd feel a lot more at home in Streaky Bay, South Australia, than Stratford-upon-Avon. From *budgie smugglers* (brief swimming trunks) to *liquid laughs* (vomit), the Australians do seem to have an awful lot of fun with their words.

When the first Europeans set foot on the island, there were perhaps 300 native Aboriginal languages. Today, that stands at around seventy, and most of them are endangered. English is the dominant language, spoken by 99.8 per cent of the population. From the moment in 1770 when Captain Cook and his botanist Sir Joseph Banks asked the local Aborigines for the name of that strange hopping animal, the Australian language has been lending, borrowing, developing and inventing words. The animal was noted down as '*gangurru*'; the myth that the *gangurru* reply in fact meant 'I don't understand you' has, alas, been debunked. The locals were simply describing a particular species of kangaroo.

Captain Cook claimed Australia for the British Crown and almost immediately this outpost on the other side of the world was turned into a penal colony. Around 160,000 male and female convicts from England and Ireland were shipped to Australia between 1788 and 1868 (when transportation ended); their numbers were swelled by the wool and gold rushes of the 1850s.

Author Kathy Lette grew up a surfer girl in Sydney but moved to London in her twenties. She points out that the Australians have spent years suffering the jibes of being a nation of convicts.

'A lot of English people see Australians as a recessive gene, sort of the Irish of the Pacific. They can't believe that they sent all the convicts out to the sun while they stayed there in the rain. My grandmother told me something fantastic when I was leaving for England, because sometimes the English can have a condescension chromosome about Australians. She said to me, "Ah Kath, you can't possibly go and live in London, that's where all those terrible convicts come from."'

Soon after the arrival of the first immigrants, a distinct accent began to emerge. One theory about the lack of lip movement in Aussie-speak is that the European arrivistes had to learn to open their mouths as little as possible when they talked to keep the flies out.

Observers described the language of the early nineteenth century as being heavily influenced by the rhyming slang of the Cockney London convicts. In 1827 Peter Cunningham, a Scottish convict ship surgeon, reported in his book *Two Years in New South Wales* that the native white Australians spoke with a distinctive accent and vocabulary: 'This is accounted for by the number of individuals from London and its vicinity . . . that have become residents in the colony and thus stamped the language of the rising generation with their unenviable peculiarity.'

Today there are three broad layers of social accents. Cultivated British English, which is spoken by around 10 per cent of the population (think actor Geoffrey Rush); a broad working-class accent (Steve Irwin); and general Australian, spoken by the majority (Kylie Minogue, Russell Crowe).

Paul Hogan in Crocodile Dundee, *1986*

There's some debate about how much of the Australian language derives from the convict immigrants and how much evolved later. The word *Pom* or *Pommie*, which the Australians use to describe the English (as in *whinging Pommie bastard*) was thought to have derived from POM – Prisoners of Her/His Majesty or Port of Melbourne, where the immigrant ships docked – or from POME – Prisoner of Mother England. Current thinking is that *Pommie* is a more recent word – as described by D. H. Lawrence in his 1923 novel *Kangaroo*.

> *Pommy is supposed to be short for pomegranate. Pomegranate, pronounced invariably pommygranate, is a near enough rhyme to immigrant, in a naturally rhyming country. Furthermore, immigrants are known in their first months, before their blood 'thins down', by their round and ruddy cheeks. So we are told.*

'Naturally rhyming country' is a most apt description, for the Australians love to play with their words: have a *Captain's* (look – from Captain James Cook); *steak and kidney* – Sydney; *dead horse* – tomato sauce. As with Cockney slang, Americans are *septic* or *seppo*. Many of the rhymes are based on popular culture. *Grundies* and *Reginalds* are undies or underpants, named after TV mogul Reg Grundy. To do a *Harold Holt* is to bolt or run away, after the Australian prime minister who disappeared while swimming. If something's a shocker, i.e. dreadfully bad, it's a *Barry Crocker* or simply a *Barry* (Barry Crocker is a popular Australian singer who sang the original *Neighbours* theme on TV).

Rhyming slang apart, some of the words and phrases are delightfully visual. Kathy Lette has made her name with her irreverent wordplay.

'I think it's something to do with our Irish heritage, because it's a love of language, there's an irreverence there, but it's often quite loquacious too.'

A beer belly becomes a *veranda*, sandwiches are a *cut lunch*, and vomit is a *technicolor yawn*. As the English reputedly don't wash, deodorant is *Pommie shower*, and if something is completely dry it's *as dry as a Pommie's bath-towel*. If you give an *Aussie salute*, you're brushing flies away; an *ankle biter* is a small child; and *don't come the raw prawn* means 'don't play the fool with me'. Someone who is mentally unbalanced has got *candles in their top hat* or *kangaroos mad in the top paddock*. A novice surfer is a *shark biscuit*, and if something's in short supply it's *scarce as rocking-horse manure*.

And then there's the unique set of diminutives Australians use – putting *ie* or *o* at the end of a shortened word. 'We shorten everything,' says Kathy Lette, 'like *cozzie, mozzie, truckie, sickie, quickie*. It's not just because it's too hot to say the whole word, it's also because it's a way of being informal and friendly. It keeps us a bit like children . . . we haven't quite grown up.'

Other popular diminutives are *arvo* – afternoon; *smoko* – tea break; *blowie* – blowfly; *sunnies* – sunglasses; *coldie* – a beer; *snag sanga* – sausage sandwich.

The Anglo-Saxon earthiness of the language reflects a deep-seated loathing of pomposity. There's a story about the Australian media mogul Rupert Murdoch, who was at a dinner party with a self-regarding broadsheet editor. The English editor announced to his fellow diners: 'I've met six British prime ministers, four French presidents, four American presidents and three popes and, do you know, not one of them struck me as having a first-class mind.' There was a pause around the dinner table, and then Murdoch said, 'Did it ever occur to you that they probably thought you were a bit of a dick too?'

It seems to be a defining quality of Australians that they can't let a remark like that go unchallenged. There's a characteristic which Australians call the Tall Poppy Syndrome – slang for someone with a big ego who needs to be brought back to the level of his peers through put-downs. It can be seen as a reasonable way of keeping inflated egos in check or evidence of an inferiority complex, a desire to punish anyone who sticks their head above the rest and is flamboyant or high-achieving or successful. The instinct to cut people down to size finds its most natural home among Australia's politicians, whose level of insult hurling can be breathtaking. Mark Latham, when he was leader of the opposition, called Prime Minister John

Howard an 'arselicker' and described the members of the Liberal Party front bench as a 'conga line of suckholes'.

The master of the colourful insult was former Prime Minister Paul Keating, whom Kathy Lette describes as 'having a black belt in tongue-fu'. Nicknamed the Lizard of Oz, he called his opponents, variously, 'gutless spivs', 'foul-mouthed grubs', 'painted, perfumed gigolos' and 'simply a shiver looking for a spine to run up'. Many of his insults were directed at John Howard, then leader of the opposition, whom he dubbed 'brain-damaged', 'mangy maggot' and 'the little desiccated coconut'. It makes Denis Healey's 'like being savaged by a dead sheep' seem quite tame.

Some of the most well known of the Australian euphemisms – the ones that have entered the British English language – flow from the pen of one particular Aussie – satirist and actor Barry Humphries. Humphries, who achieved worldwide fame with his alter ego, Dame Edna Everage, travelled to London in the 1960s and wrote a comic strip for the satirical magazine *Private Eye*. 'The Adventures of Barry McKenzie', illustrated by Nicholas Garland, chronicled the exploits in London of Bazza, an uncouth, loud-mouthed, beer-swilling 'ocker'. Bazza, writes Humphries, initiated readers 'into the mysteries of Australian colloquial speech'. He spoke a 'synthetic Australian compounded of schoolboy, Service, old-fashioned proletarian and even made-up slang'.

In fact, Humphries created so many of his own made-up euphemisms that have entered the vernacular that it's almost impossible to say what's invented and what isn't. Most of the expressions relate to bodily functions and sex. *Multicoloured yawn, pointing Percy at the porcelain, siphoning the python, one-eyed trouser snake, dining at the Y, shaking hands with the wife's best friend, sinking the sausage.* Some phrases, like *chunder* (to vomit) and *up shit creek*, were dying out until Humphries resurrected them.

Political satirist John Clarke explains: 'Barry is hugely observant but he's much more creative than your normal observer. So what he's done is enriched this series of observations by making the story better than he actually heard. So he's a Shakespeare, he's added to the language.'

The comic strip was banned in Australia as, according to Customs

Barry Humphries and Barry Crocker in The Adventures of Barry McKenzie, *1972*

and Excise, it 'relied on indecency for its humour'. Subsequent made-in-Australia feature films based on the book – with Barry 'Shocker' Crocker playing the lead – were, however, supported by the Australian government. In fact, the prime minster, Gough Whitlam, made an appearance in the 1974 film *Barry McKenzie Holds His Own*, where he granted a damehood to McKenzie's aunt, Edna Everage.

Barry Humphries and a new generation of Australian writers and comics continue to keep the language alive with colourful imagery, but there is a sense that – bit by bit – Australian slang is under threat.

John Clarke agrees.

'It's very seldom that someone like Barry Humphries comes along, and I think somebody who can infuse the language with an enormous amount of imaginative metaphor and imagery that people will pick up is pretty unusual. My impressions of young people is that they are using the language of the internet and the social networking sites a lot more, and that language is a kind of shorthand in some respects.'

It's a case of Save our Slang, says Kathy Lette.

'The American influence is huge, because all the kids are watching the American programmes. So they've suddenly started talking like they're in the New York ghetto – "see you later, dude" and all that stuff. It's a real push to protect our slang, because we've actually realized that it is something quite precious and colourful and historic.'

I  Slang

Thousands of new words are added to the English language every year. No one knows just how many, but according to the word-tracking Global Language Monitor, a new word is created every ninety-eight minutes. It's an extraordinary thought, especially since whole minority languages are disappearing at a similar breathtaking pace. What's undoubtedly true is that the influence of television and films, computers and social networking in the last decade has meant

The first graphical symbol to enter the Oxford English Dictionary

the greatest explosion of vocabulary since Shakespeare. *Texting, Twitter, blog, memory stick, download, carbon footprint, 24/7* and *9/11, ground zero, bling, chav, credit crunch* . . . there's a seemingly endless list of words which we use on a daily basis today which simply did not exist a decade or so ago. Our language is evolving and expanding on a global scale like never before.

New entries to the 2011 online edition of the *Oxford English Dictionary* include: ♥ to heart – meaning to love (e.g. I ♥ New York) – the first graphical symbol in the OED's history; *cream-crackered* – rhyming slang for knackered, i.e. exhausted, *lashed* for drunk, *fnarr fnarr* for a lecherous snigger, *dot-bomb* meaning a failed internet company and *couch surfing*, 'the practice of spending the night on other people's couches in lieu of permanent housing'. *Tragic* has a new, twenty-first-century meaning: 'a boring or socially inept person, especially a person who pursues a solitary interest with obsessive dedication'. The new entry *wag* is defined as an acronym for 'the wives and girlfriends of any group of men, especially celebrities or sportsmen', first mentioned in a *Sunday Telegraph* report in 2002 about the England football team's partners; according to the *OED*, '*Wag* is notable for the extremely fast journey from its introductions to the language to its use as usual English vocabulary.' *Muffin top* enters the dictionary as 'a protuberance of flesh about the waistband of a tight pair of trousers'. Although many of the words are computer and social networking jargon, television has influenced our everyday language as well. One TV programme in particular, *The Simpsons*, has created a raft of popular words and phrases. This hugely popular animated series about a dysfunctional American family has been on air since 1989 and, according to Mark Libermann, director of the Linguistic Data Consortium at the University of Pennsylvania, 'has apparently taken over from Shakespeare and the Bible as our culture's greatest source of idioms, catchphrases and sundry other textual allusions'.

D'oh, the grunt uttered by Homer Simpson, is the cartoon's most popular neologism. *Doh* without the apostrophe entered the *OED* in 2001 as 'expressing frustrations at the realisation that things have turned out badly or not as planned or that one has just said or done something foolish'. *Meh*, meaning 'whatever' or 'boring', is another

Is Homer Simpson the new Shakespeare?

Simpsons expression that has become particularly popular in the lexicon of web conversations. The *Collins English Dictionary* included it in 2008, giving as an example 'The Canadian election was so *meh*'; it's not in the *OED* – the dictionary's new words editor says it hasn't quite yet passed into 'widespread unselfconscious usage' – but he's keeping a 'meh' file. Other popular expressions include *lupper* (a calorie-laden meal in between lunch and supper) and the insult *eat my shorts!* In one episode the writers included two nonsense words, *embiggen* – to enlarge – and *cromulent*, meaning fine. 'He's embiggened that role with his cromulent performance,' says Principal Skinner. Both words were so convincing that *cromulent* has been included in various slang dictionaries and *embiggen* was used in a scientific paper on string theory: 'there is a competing effect which can overcome the desire of the antiD3s to embiggen, namely their attraction towards the wrapped D5s'. Most cromulent.

Textese

The mobile phone and Facebook have become so much a part of personal communication that a whole new shorthand language has developed. Purists dismiss it as a sort of viral disease of *GR8s* and *4Us* which is wrecking language with its lazy spellings and impenetrable acronyms. Yet language needs to be expedient, and abbreviations or acronyms can be jolly useful. We happily talk about *ASAP* or *DIY* or *BYOB* (bring your own bottle). So the texter or the Twitterer or the Facebooker, who needs to be speedy and save money by using as few letters as possible and avoid finger strain, abbreviates words – often in highly ingenuous ways.

Some of these abbreviations are fairly functional: *GR8* (great), *BRB* (be right back), *ATB* (all the best) and *HRU* (how are you?). Some bits of textese have become so widespread that, according to the *Oxford English Dictionary,* they've officially entered the English language. New entries for 2011 are *OMG* (Oh my God) and *LOL* (laughing out loud), joining *IMHO* (in my humble opinion), *TMI* (too much information) and *BFF* (best friends forever). Textese can be playful, personal – *MWAH* to suggest the sound of sending a kiss and *RME* for 'rolling my eyes'. The really interesting words are those which are stepping out of the text zone and entering spoken language. *LOL* is now used as a noun for laugh – *lol* – or a verb – *to lol*. *JK* for 'just joking' is added at the end a sentence: 'I forgot your birthday . . . JK!'. *Soz!* has become a lighthearted 'sorry'.

Keypad symbols can portray moods in texts – although they do require a certain repositioning of the mobile phone to get the full effect:

:) (happy),
:((sad)
:-D (laughing)

If you want to describe your reaction to something funny, you might send a text combining symbol and abbreviation :)E2E, meaning 'grinning ear to ear'.

Textese isn't limited to youth – after all, everyone is looking for ways to send messages as quickly and cheaply as possible. A competition for a texting Poet Laureate awarded second prize to this love poem:

> *O hart tht sorz*
> *My luv adorz*
> *He mAks me liv*
> *He mAks me giv*
> *Myslf 2 him*
> *As my luv porz*

The ode to her husband was texted by a sixty-eight-year-old grand-mother from Lancashire.

Teenglish (or Romeo and His Fit Bitch Jules)

What's fascinating about the way language evolves is the increasing influence young people have on it. Just as from the 1960s onwards a youth culture has dominated music and fashion, so playfulness and experimentation in language has become the preserve of the young. In today's multicultural Britain, the language of the school and the street is influenced by West African, Afro-Caribbean, Asian and black American as well as urban British. Take the word *bling* (or sometimes *bling-bling*), which originated in American hip-hop culture to mean ostentatious clothing or jewellery, possibly imitating the clashing sound of jewellery or light reflecting off it. *Bling* was taken up by young people in the UK at the beginning of the twenty-first century, adopted by an older generation within a few years and is being dropped now as old hat by the younger generation. Or *nang*, a word meaning excellent or great, which seems to have spread in the last decade from Bangladeshi communities in the East End of London to other ethnic groups. It's thought to be from a Bengali expression for a naked woman.

Ali G: 'voice of da yoof'

The opening line of *Romeo and Juliet* was translated into London slang by satirical author Martin Baum in *Romeo and His Fit Bitch Jules*: 'Verona was de turf of de feuding Montagues and de Capulet families. And coz they was always brawling and stuff, de prince of Verona told them to cool it or else they was gonna get well mashed if they carried on larging it with each other.' Baum, like the fictional TV character Ali G (hip-hop obsessed 'voice of da yoof'), uses Jafaican, a mixture of Jamaican, Asian and Cockney. Linguists refer to it as multicultural London English, a patios which is increasingly the language of young, inner-London, working-class people and has introduced words like *bare* for 'very' ('I'm bare hungry'), *peng* for 'attractive' or 'hot', *yoot* for child or children and *yard* for home.

Many of these words spread out from London to the rest of the country via television or music or Facebook, although a recent unscientific but revealing BBC survey of teenage slang around Britain showed huge regional variations. Teenagers were given the following sentences to translate into their own local slang:

John's girlfriend is really pretty. But she got mad with him the other day because he wanted to hang out with his friends rather than take her to the cinema. She got really angry and stormed off. It was very funny.

Teenagers from a school in west London put it into their own street slang:

John's chick is proper buff but she switched on her man the other day cos he wanted to jam with his bred'rins instead of taking her out to the cinema. She was proper vexed and dust out. It was bare jokes.

Compare the London slang with that of a school in Keighley, West Yorkshire:

Jonny's bird is proper fit and she got in a right beef the other day cos he'd rather chill with his mates than go to the cinema. She got stressed and did one. It was quality haha.

And a school in Swansea in Wales:

John's missus is flat out bangin'. But she was tampin' the other day 'cause he bombed her out for the boys instead of going to the cinema. She started mouthing. It was hilarious.

Use of the word *innit* – probably British Asian in origin – is increasingly commonplace amongst young people. In its simplest form it means 'isn't it?' – 'That's right, innit?' – but its usage is being expanded in a rather interesting way. In linguistics, phrases like 'isn't it?' at the end of statements are called tag questions. English has lots of them. 'You'd better go now, hadn't you?' 'Oh I must, must I?' 'That's quite a wind blowing, isn't it?' Other languages often have just one fixed phrase or invariant tag. In French, it's *n'est-ce pas?*, in German, *nicht wahr?*, in Spanish, *¿verdad? Innit*, according to some linguists, is being used by young people as an invariant tag and may end up providing English with the *n'est-ce pas* it lacks. It's quite a thought, innit?

Ins and Outs

Slang and jargon serve a similar purpose – to separate the 'in' from the 'out' group. And nowhere are these definitions more important than in school. Berkeley High is a state school in the San Francisco Bay Area with over 3,000 students. It's unusual in having a large, educated, middle-class catchment mainly from Berkeley University as well as many lower-paid immigrant families. It's a melting pot of races – white, Afro-American, Latino, Asian and lots of mixes in between; a veritable Tower of Babel. So the desire to establish a common but exclusive slang vocabulary is strong. The students even compiled their own dictionary of slang, which, when published in 2004, became a local bestseller. Much of the language comes from African-American hip-hop, but Chicano (Mexican English), Jewish, Hindi, punk and sports cultures also contribute words and expressions.

One of the students, Connor, explains: 'Berkeley is one of the most diverse places you'll ever be. We have so much slang because we have every ethnic group, we have every social class. It's the only public school in Berkeley, so everybody that's not in a private school is here in Berkeley High. So the slang is different, and that's why there's so much of it . . . We have all the language and we use it all.'

The current students dismiss the 2004 dictionary as being already outdated; school slang has moved on. One girl describes with great excitement how words change overnight; a random expression appears, and suddenly everyone's using it.

'It can change in one day. Like one hour to the next, it can be just a new word . . . Someone can know someone and says, "Oh this isn't new," and then someone else says, "I've never heard that before in my life." That's how weird it is because it's such a huge range of people.'

'But eventually it will become nationwide.'

'Yeah, it becomes big. So it just travels.'

'I know people who go to St Mary's [College High School in Albany, California], and I was with them when they started a word and then like a month later I heard it at Berkeley High. So they made up the word, they spread it around St Mary's, someone hung out with someone at St Mary's, heard it and started saying it here.'

Changing schools can be like moving to a foreign country, 'literally like trying to learn a new language'.

Social networking – Facebook and Twitter – can speed up the process: slang can spread like a disease.

'Yeah, but slang's not a disease!' shouts one.

'It's a positive thing,' says another. 'It's like language is gonna evolve one way or another.'

'Language does not have a right and wrong. Language is just how people communicate.'

These teenagers derive real pleasure from their language, constantly creating or borrowing new words, rejecting old ones. The students give a quick run-down on current school slang.

'When some people say you're cool, they say, "you got swag".'

'If I say "swag points", it means that you just said something hella cool.'

Hella puts emphasis on everything – that girl's *hella* pretty. *Hella* has become so much part of North Californian vocabulary that scientists have been lobbying for it to join the likes of *mega* and *giga* as an internationally accepted prefix for a number with twenty-seven zeros.

Arkotalko, apparently, is a brand new word for awkward. A *scrub* is a loser and a *cuddy* is a friend, as is *cuzzo*, *homie* and *dog*. And if you go *hypi*, you go crazy.

The students talk about the role played by music and dance moves in the spread of slang words.

'There are certain words used in certain areas because there are certain dances and certain movements going off in different areas. Like when you say a "hyfi" in Oakland, they call it "the hyfi woman" and then in LA, it was a dance called "jerking". And that was a major thing in LA. People still did it in different areas but it's major areas where things start and where they're most popular.'

'It can start anywhere. Even if it's started in some weird town in Ohio, it's all about the song. 'Cos when you put a song on iTunes, everyone in the world can hear it. So that's kind of just how it spreads.'

Hip-hop and Rap

Pop culture from the early twentieth century onwards has injected countless new words and expressions into our language. As the students from Berkeley High point out, music acts as a vector for words; one of the most influential forms these days is hip-hop and rap.

Hip-hop is a style of music which started in the 1970s in the United States amongst the African-American and Jamaican-American communities of the big cities. It began with disc jockeys creating rhythmic beats by repeating small portions of songs on two turntables. This was later accompanied by rap, which is the lyrical part of hip hop, a sort of verbal rhyming chanted over the beats. Rapping is also called *MCing* or *emceeing*, short for Master of Ceremonies.

H. Samy Alim, a professor of linguistic anthropology at California's Stanford University, specializes in black language and hip-hop culture. In the company of rapping DJ Kenard 'K2' Karter, he discusses how rapping can be seen as an extension of an African-American oral tradition that stretches back through the passionate oratory of the black preachers to the generation-to-generation storytelling of the slaves.

'The word rap,' says Professor Alim, 'was used in the black community long before it was associated with music. If you could rap, that was your talking ability . . . The gift of the gab.'

'It's metaphors,' adds 'K2'. 'The hip-hoppers we see today aren't just describing their own experiences; they're describing the experiences of others.'

Professor Alim is excited by the linguistic daring of rap: 'The MCing or rapping is a verbal art form that depends on your delivery, your lyrical inventiveness, your ability to create new rhyming structures. It's something that really speaks the truth to you – a punchline that could be funny and makes people want to rewind and hear it again or a story that's really powerful and moving and gives you goose bumps and the hair on your neck stands up.'

Rappers have different styles of what's called 'flow', the rhymes and rhythms of their lyrics, with names like 'The Chant' and 'The

Hip-hop, one of the most influential music forms

Syncopated Bounce' and 'Straight Forward'. Staying on the beat is crucial as, of course, is not stumbling mid-flow; the smallest falter would ruin the whole effect. For rapping to work, it's got to be perfect, and as in most things practice makes perfect. So how does a rapper practise his art? Clearly not alone in the bedroom.

'You step up into the cipher,' explains Professor Alim. 'A cipher is like this: we're having a little cipher right now, we've got a circular group. We're talking together, we're building together. As an MC you're in a cipher of MCs, you're building on each other, sharpening your skills. It's like a lyrical testing ground, a battling ground, a stomping ground.'

Ragtime, jazz, rhythm and blues, rock and roll, soul – African-American music has always crossed over into the broader American culture, bringing with it the language of *cat* and *hip* and *funky* and *hot* and *cool* and *chill*. Clearly hip-hop has travelled the same route, its music and slang adopted by people throughout American society. Professor Alim gives an example of how a phrase which seems to start in the music world gets taken up by the broader culture and then gets degraded.

'I was in the locker room, working out in the gym, and there's a white gentleman of about sixty talking on his cellphone. He ended his conversation and he said, "All right, hit me back later." And I thought, "What? Since when . . . ?"'

It's an example of cultural appropriation which these rap experts have mixed feelings about. Here's 'K2':

'From a language perspective, hip-hop becomes the reward for describing social linguistic differences in culture or describing the experience. It's the modern-day expression for saying, "It's okay for us to have differences and we can share our experiences, and I don't have to totally get it."'

'And that,' adds Professor Alim, 'is what some people view as cultural theft. Where it's a positive borrowing, it could be a building of relationships across racial lines. But it can also be viewed within a context of racial discrimination that goes back decades and decades – I can borrow your language but I'm sure as hell not going to borrow your experience.'

Professor Alim thinks that rap's short, to-the-point sentences make it an ideal language for Twitter and Facebook. It's one of the reasons why, from Tokyo to Timbuktu, rap has such widespread youth appeal and why, in 2011, this verbal art form is having such a profound affect on social movements around the world.

Rapping and Revolution

A rap by a twenty-one-year-old is thought to have helped bring down the Arab dictatorship of Tunisia. Hamada Ben Amor – aka El General – was arrested after he recorded and uploaded a rap protest song on to Facebook. 'Rais Le Bled' ('President, Your People') is an extraordinarily brave personal message to the now former President Ben Ali: 'My president, your country is dead / People eat garbage / Look at what is happening / Misery everywhere / Nowhere to sleep / I'm speaking for the people who suffer / Ground under feet.'

Crucially, it was recorded in Arabic, so it spread like wildfire on the social networks, from Casablanca to Cairo and beyond. Ben Amor's arrest seems to have inflamed the protests, with even more young Tunisians taking to the streets. Within a week, Ben Amor had been released, President Ben Ali had fled the country and the protest rap was being listened to throughout the Arab region.

As another Tunisian rapper, Balti, commented after the overthrow of the old regime: 'The revolution is a social movement, and rap is always talking about social issues. We come from very tough neighbourhoods and we talk in our songs about social problems such as unemployment. We feel like our voices didn't get to the regime, to the top officials, but thank God our voices were heard by the people, so we were the fuel of our revolution.'

El General, Ben Amor, emerged from jail a star. He immediately put out a new rap – an 'ode to Arab revolution' – with the lyrics 'Egypt, Algeria, Libya, Morocco, all must be liberated / Long live free Tunisia.'

As El General proves, language is a powerful tool. It should be used with care but it shouldn't be restricted. In fact it's impossible to put a lid on language, and, by making words or expressions taboo, you give them even more power. Language should be left to grow and evolve into more colourful and creative forms for future generations.

CHAPTER 4

Spreading the Word

Writing is, quite simply, the greatest invention in human history – in fact, it made history possible. 'Without words, without writing and without books,' said the German author Hermann Hesse, 'there would be no history, there would be no concept of humanity.' Or, as H. G. Wells put it, 'Writing put agreements, laws, commandments on record. It made the growth of states larger than the old city states possible . . . The command of priest or king could go far beyond his sight and voice and could survive his death.' We humans have been around for nearly 200,000 years, but we've only had writing for a little over five millennia. It's taken us that long to develop from pictures and scratches on clay and parchment to emailing on a laptop computer. Different writing systems evolved gradually over long periods of time, and there's still debate about which was the very first. The British Museum has thousands of objects with ancient writing on them, some as old as 5,000 years. Some historians say the hieroglyphs of Egypt have the edge, but Assyriologists like Irving Finkel, who study the ancient Mesopotamian culture, claim that its writing form – called cuneiform – is the oldest.

Cuneiform

Mesopotamia, which means the land 'between the rivers', was the region around the Tigris and Euphrates rivers, largely forming modern-day Iraq and parts of Syria, Turkey and Iran. Around the third millennium BC, Mesopotamia was divided into the Sumerian civilization in the south and the Akkadian in the north. It was the Sumerians who invented the fascinating *cuneiform* writing system that looks like a series of spindly bird tracks. The scribes used a blunt reed, cut at the tip into a wedge shape, which was pressed down into soft clay in various different ways to cut vertical, horizontal and oblique lines. 'Cuneiform' comes from the Latin *cuneus*, meaning 'wedge'.

A cuneiform inscription dating from 2100 to 2000 BC

The cuneiform writing system was used for more than 3,000 years. It developed from pictographs, which were realistic drawings to communicate an idea, into a syllabary form, where signs stood for sounds as well as words. For example, says Finkel, initially there would be a sign, a picture representing, say, 'beer', and every time you wanted to say 'beer', you would draw that picture. But later on came a gigantic leap in the journey of how writing developed: the picture sign came to stand, not for what it looked like, but for the *sound* that sign represented. What began as symbolic drawing evolved into a system which could be used to express a language phonetically. So, in the case of 'beer', a sign could stand for any word that sounded like *beer*, or had a *b* sound, and it could be varied and developed, and soon you had a single unit of expression of sound, rather than a slightly more clumsy series of pictures. That gave flexibility and adaptability, and nuance, so that instead of just being able to say, 'three bottles of milk' and drawing the sign three times, you could say, 'Please leave three milk bottles on Sunday morning. Last week it was off.'

Why did people want to have a written form of communication in the first place? On the simplest level it was for making lists, for counting, for keeping a record in areas like trading and tax and money. Mesopotamia was largely a population of shepherds and farmers, and archaeologists have found early clay tablets inscribed with lists of sacks of grain and heads of cattle. Later tablets contain more complex information about social structures, e.g. what sort of jobs people did and how financial dealings were conducted; or they preserve religious texts and stories like the *Epic of Gilgamesh*, which tells of a legendary Sumerian king's quest for immortality.

We've even unearthed the ancient equivalent of school jotters – tablets found in Sumerian schools which show the teacher's text on one side, and the pupil's copy on the other, so we know how people learned to write cuneiform. It wasn't an easy task: the ancient Mesopotamians had to master the art of inscribing hundreds of signs and learn the various meanings that they might have, depending on the context. It was an art confined to a small number of scribes, who became an elite group in society, more powerful sometimes than the illiterate courtiers or even the king himself. Knowing how to write

Recording the allocation of beer. An upright jar with a pointed base appears three times on this tablet

was a source of power, and a privilege; it took years to learn and was never a system designed for the masses.

Cuneiform expert Irving Finkel has a special chart with the English alphabet and their cuneiform signs. He teaches children cuneiform by getting them to write their name down with a pencil, divide it up into syllables and then find the syllables on the chart with which their name can be spelt. That's not simple, because Far Eastern languages have a very specific pool of sounds, and consonants can't be written on their own, only as vowel-consonant or consonant-vowel.

The fact that cuneiform documents were written in clay has meant that vital tablets have survived for our examination. When ancient libraries or other buildings housing the tablets were burned, the clay was fired and so preserved for posterity. The British Museum has a staggering 130,000 clay tablets in its safe keeping, and there are probably millions more still buried under the sands of Iraq and Syria. That means, Finkel adds casually, there's about three centuries of work ahead for Assyriologists like him as they continue to learn from these records of humanity's past.

Professor Finkel is clearly very excited by the vision of the treasures out there but is there actually something fundamental about our species that can be learned from the study of these ancient peoples and their writings? Oh yes, he says. To him, the people who scratched on clay with a reed and told us about their economy and culture and stories are basically just like us. 'They were afraid of disease and impotence and not having children and having no money and warfare like everybody else,' he says, 'and they told lies and they were truthful and they were committed and they were hypocritical.' He believes we'd know them on the Underground, these ancient people. 'The whole dazzling range and wonder of the human mind can be plucked out of these tablets. They were written by people like us.'

The Decipherers

The clever and crucial thing about cuneiform writing was that it could be adapted to write languages other than Sumerian or Akkadian, like Babylonian, Hittite and Old Persian. Between the third and first millennia BC it travelled as far south as Palestine and as far north as Armenia. And wherever it went, it left its words on plaques, clay tablets and ancient stone walls – to be discovered, puzzled over and eventually deciphered thousands of years later. That's the exciting part of the story. It was thanks to a few brilliant, dedicated men that the secrets of these written languages were unlocked and vast areas of our ancient history reborn.

The French academic (and coiner of the word *franglais*) René Etiemble was full of admiration for the decipherers who wrestled with the messages from the past: 'Let us ask ourselves, positively,

Georg Friedrich Grotefend, one of the greatest decipherers

flatly, whether perhaps we should not admire those who deciphered hieroglyphs, cuneiform or Cretan Linear B, a little more . . . than those who designed the first pictograms, or who created a system for representing a complete vocabulary with a few alphabetic signs.' These were men who made it their life's mission to interpret the obscure signs that were engraved on stone, painted on walls and tombs or scratched in clay. They were puzzle-solvers, detectives, archaeologists and linguists rolled into one. While Jean-François Champollion was working on his first translation of the Rosetta Stone hieroglyphs in 1822, another group of scholars was making momentous discoveries about cuneiform.

It was a piece of inspired guesswork by a scholar and schoolteacher from Germany that led the way. Georg Friedrich Grotefend was an accomplished Latinist and linguist, but he had no background in Oriental languages, so it's all the more extraordinary that in 1802 he succeeded in partially deciphering the old Persian cuneiform writing and provided the foundation for later work to provide a complete translation of the signs.

Archaeologists had already started collecting artefacts from ancient Assyria and Babylonia and had noted the strange script on seals and clay cylinders, and on the walls of the ruined palaces of the Persian kings at Persepolis. Various travellers and scholars, as far back as the seventeenth century, had copied and written about these mysterious scripts, so people knew about them – they just couldn't understand them. Then the Danish government sent the German mathematician and cartographer Carsten Niebuhr on a scientific exploration of Egypt, Syria and Arabia. On his way back, Carsten made detailed and laborious copies of the inscriptions at Persepolis – sacrificing his sight in the process – and in the early 1780s published three volumes of his work. At last, scholars in Europe had extensive and accurate records to use, and the work began in earnest.

By the time Grotefend came along scholars had recognized there were three different forms of the script at Persopolis and that the systems were alphabetic. It's a bit like a scholarly detective story, this quest to decipher and understand. It was all about finding clues and trying to fill gaps, comparing notes and building on other people's theories until the moment when things clicked into place. Grotefend looked at the three different scripts and decided that they represented three different languages. Here was a powerful king communicating his edict or story in more than one language so that it could be understood across an empire. He also logically worked out that the first script must be in old Persian, the language of the kings, and the other two must be translations. He was feeling his way through deduction, common sense, available information and his linguistic training.

It was like trying to solve the most nightmarishly difficult *Times* crossword you can imagine. He picked out repetitive phrases, which were used to honour Persian kings. He then compared those letters with the kings' names, which he knew from Greek historical texts, followed the pattern of inscriptions in Middle Persian which linguists had recently deciphered and eventually worked out the approximate phonic values of about ten characters.

The sting in the tail to Grotefend's story is that the academic world turned up its nose when his paper was presented to the Göttingen Academy of Sciences. They didn't understand the inductive method he'd used and, frankly, they didn't believe a schoolteacher who specialized in Latin and Italian and hadn't studied Oriental languages. It would be another thirteen years before a friend eventually published the paper.

Other scholars then built on Grotefend's work and in 1905 François Thureau-Dangin produced the first translation of Sumerian, man's earliest identified writing and the original version of cuneiform – which was where we started this chapter.

The Rosetta Stone

We don't know whether writing was discovered or invented independently in different parts of the world, or whether it spread by osmosis. But by 1200 BC writing was being used in India, China, Europe and, of course, Egypt. Some scripts, like Cretan Linear A, are still a mystery, but we did find the key to the words of the pharaohs, these beautiful, exotic hieroglyphics which cover countless monuments and tombs, thanks to the Rosetta Stone.

The Rosetta Stone is a big lump of dark granite 45 inches by 28 inches by 22 inches, on which there are three inscriptions. The upper text is in hieroglyphs, the middle portion is demotic script (the language which came between Late Egyptian and Coptic), and the lowest is in classical Greek. Because it presents essentially the same text in all three scripts, scholars could compare the Greek words, which they knew, with the hieroglyphs and their knowledge of later Coptic, and so learn to read the signs.

The stone was discovered during Napoleon Bonaparte's expedition to Egypt in 1799 and seized by the British when they defeated the French in Alexandria in 1801. It dates from the second century, when the priests who had gathered at Memphis to celebrate the arrival of the young Ptolemy V composed a decree in his honour which was copied on to stone, with the two translations. The stone was placed in the British Museum in 1802 and it's been there ever since.

Jean-François Champollion deciphered the first hieroglyphs on the Rosetta Stone

The Rosetta Stone, divided into hieroglyphs, demotic script and classical Greek

It was a brilliant young French scholar and philologist called Jean-François Champollion who first deciphered the hieroglyphs on the stone and demonstrated that the Egyptian writing system was a combination of phonetic and ideographic signs. Like so many of the other code-breakers of the ancient writings, he was a superb linguist with a genius for inspired thinking and deduction. He was also obsessed, a workaholic who devoted his life to his studies and died at the age of forty-two. He first heard of the Rosetta Stone as a boy, and it fired an interest in hieroglyphics which burned fiercely throughout his student days in Grenoble. There he studied a multitude of languages, including Latin, Greek, Sanskrit and Coptic, and became convinced that Coptic was simply a late form of the language which the ancient Egyptians had spoken.

Champollion spent three years working on deciphering the hieroglyphs on the Rosetta Stone, and his monumental contribution to history was to prove that hieroglyphs were not simply pictures, but that they represented sounds as well. His 1824 work *Précis du système hiéroglyphique* gave birth to the modern field of Egyptology.

Michael Ventris

In 1936 the famous archaeologist Sir Arthur Evans, who had excavated the ancient palace of Knossos in Crete, held a conference in London on the mysterious writing system inscribed on the clay tablets which he had uncovered, a script he had spent decades trying to decipher, without success.

There was a young schoolboy in the audience that day, who was thrilled and inspired by the story of a secret code which no one could break. His name was Michael Ventris and sixteen years later, on 1 July 1952, he made a BBC radio broadcast which was to secure his place in archaeological and history books for ever. He announced that he'd deciphered Linear B, Europe's earliest known, and previously incomprehensible, writing system.

Michael wasn't an archaeologist or scholar, he was an architect, but he was fluent in several languages, and his life's passion was cracking the Linear B code. He was a loner, an outsider, an unconventional thinker from an unconventional background who'd been brought up by his divorced Polish mother and educated for a time in Switzerland.

Throughout his years of architectural training and practice, which was interrupted by the war, in which he served as an RAF navigator, Ventris worried away at Linear B. He worked on the problem through a series of small, detailed steps, helped by the researches and insights of others, like the American classical scholar and archaeologist Alice Kober, who observed that certain words in Linear B inscriptions had changing word endings – perhaps declensions like Latin or Greek.

Ventris took this clue and painstakingly constructed a series of grids mapping out the symbols on the tablets and associating them with consonants and vowels. He still didn't know *which* consonants and vowels they were, but he had learned enough about the structure of the underlying language to begin guessing.

Linear B tablets had also been discovered on the Greek mainland, and Ventris believed that some of the chains of symbols on the Cretan tablets were names. But certain names appeared *only* in the Cretan texts, so Ventris made the inspired guess that those names applied to cities on the island. This proved to be correct. This gave him a set of symbols he could decipher from. He soon unlocked much of the text and demonstrated beyond any doubt that Linear B was the writing system used by the Mycenaean inhabitants of continental Greece, and that the underlying language was an archaic dialect of Greek, centuries older than anything previously known.

Instant celebrity followed, and in 1955 Ventris published, with the philologist John Chadwick, the monograph *Documents in Mycenaean Greek*. The following year, at the age of thirty-four, he was dead, killed in a car crash. John Chadwick said of the particular genius of Michael Ventris: 'Ventris was able to see, in the confusing diversity of these signs, an overall pattern and to pinpoint certain constants which revealed the

	a	e	i	o	u
d					
j					
k					
m					
n					
p					
q					
r					
s					
t					
w					
z					

Linear B is based on combinations of sounds

Michael Ventris cracked the Linear B code

underlying structure. It was this quality – the gift of being able to make order out of apparent confusion – that is the sign of greatness amongst the scholars in this field.'

The Alphabet

By 1200 BC, different writing forms were being used in India, China, Europe and Egypt by 1200 BC. One feature that cuneiform, hiero-glyphs and Chinese characters shared was that all three could be used to transcribe either words or syllables. Both the Chinese writing system, which developed around 2000 BC, and the Egyptian hieroglyphs – which means 'writing of the gods' – are exceedingly beautiful, functional as well as artistic. They're complex, too, because they're made up of several different kinds of signs.

Hieroglyphs comprised pictograms, phonograms and determinatives – signs used to indicate which category a sign is in. In Chinese a single sound can represent several things, depending on how it's written. For example, the sound *shi* can mean 'to know', 'power', 'world', 'to love', and many other things, according to the other elements with which it's combined. Each character is formed of a key, which gives the basic meaning, and a phonetic element to indicate pronunciation. It doesn't stop there, though. The calligraphy of Chinese – the forming of the characters – also contributes to meaning, through the style of the writing, the colour of the ink and the intensity of the stroke. So, to varying degrees, the systems of cuneiform, hieroglyphs and Chinese meant that you had to master a large number of signs and characters to read and write them.

Then, a thousand years before the birth of Christ, a massive change took place in the history of writing: the alphabet was invented. Not overnight, obviously, but gradually in different places, over a long period of development, crossover and osmosis.

An alphabet works differently from the older systems. The key difference was the inspiration that, once you have a set of symbols that represent *sounds* and not things, you can represent any noise, any word in any language, using the same symbols. In principle, it's possible to write almost anything with about thirty letters. More or less. We've got twenty-six letters, but that's not enough for us to transcribe all the sounds in English, which is why our spelling is more difficult to learn than, say, Italian.

English has what's called a deep orthography – there's no one-to-one correspondence between the sounds (phonemes) and the letters (graphemes) that represent them. It's notorious for its inconsistencies, for instance in words where the same letters are pronounced differently like *cough*, *bough*, *dough* and *tough*; or *chat*, *chute* and *scholar*. On the other hand, sometimes different letters are pronounced the same way, like the *sh* sound in *ship*, *initial* and *machine*. So learning to spell and write English can be difficult, but twenty-six letters is nothing like the 1,000 basic signs needed for writing Chinese, the hundreds of hieroglyphs used in

Egyptian, or the 600 cuneiform signs that were hammered for six years into students in the scribe schools in Mesopotamia. With an alphabet came the possibility of spreading learning out from the elite few to the many.

So where did our alphabet come from? It's a story of twists and turns and many scenes. Like so much of our culture, our letters are an overseas import, brought to these shores by the Romans. But it's not as simple as that. The Romans didn't invent the alphabet either – they had adopted the Etruscans' writing system, which they in turn had borrowed and adapted from the Greeks. We get our word *alphabet* from their two first letters – *alpha* and *beta*. But where did the Greeks get their alphabet from? Some romantically minded scholars have proposed that a brilliant contemporary of Homer was inspired to invent the alphabet to record the poet's oral epics, the *Iliad* and the *Odyssey*. It seems unlikely, but Homer does give us a clue about the origins of writing. In the *Iliad* and the *Odyssey*, he mentions the Phoenicians, traders who travelled the Mediterranean on ships.

Another Greek myth also points to the Phoenicians. Herodotus claimed that the alphabet was brought to Greece by Cadmus, the legendary founder of the Phoenician city of Thebes. The Phoenicians came from the coastal areas of present-day Israel and Lebanon and part of Egypt. Their cities were places like Tyre, Sidon and Beirut.

'Phoenician' means 'dealer in purple'; one of their most important items of merchandise was the purple dye extracted from a large type of sea snail. We call them Phoenicians, but they called themselves Kenaani or Canaanites – as in the Canaan of the Bible, the land flowing with milk and honey.

Phoenician was the first widely used script in which one sound was represented by one symbol, and it's been called the Mother of Modern Writing. The Aramaic alphabet, a modified form of Phoenician, was the ancestor of the modern Arabic and Hebrew scripts. The Greek alphabet, and its descendants – Latin, Cyrillic and Coptic – was a direct successor of Phoenician, though the Greeks changed certain letters to represent vowels. The big move the Phoenicians made was turning a symbol for a thing into a symbol for a sound. For example, in the case of the letter *aleph* which was formed in

PHENICIAN	ANCIENT GREEK	LATER GREEK	ROMAN

Comparison of ancient Phoenician, ancient Greek, later Greek and Roman Alphabets the shape of a head of an ox, the Phoenicians transformed it from a representative symbol to an arbitrary mark that didn't have to resemble an ox head any more, and stood for a sound, something like *aak*.

We don't have many surviving examples because the Phoenicians wrote on papyrus and shells, not on tough, solid clay which was preserved by firing. They didn't seal their scrolls in jars like the Egyptians or preserve them in tombs, so much has disintegrated in the salty sea air of the Mediterranean. However, they did mark their possessions and objects they made with names and inscriptions, and later also to commemorate the dead on tombstones, and we do have fragments dating back to the ninth century BC which show this non-cuneiform alphabet.

The letters themselves are angular and rough-looking, and the alphabet only contained consonants, rather in the way other Semitic languages like Hebrew and Arabic are often written without vowels. The twenty-one letters were usually written from right to left and incised with a stylus.

The Phoenicians were a tremendously successful maritime culture, the trading kings of the Mediterranean, and they moved around, making lots of money and contact with different cultures from north Africa to Greece. And as they travelled and traded, they spread this new writing system, knocking down the old barriers of an elite writing class. The Assyrians had their form of writing, and the Egyptians theirs, but they kept them to a priestly caste, a sort of scribal closed shop. The Phoenician alphabet was much easier to learn and it was widely disseminated among all sorts of people and cultures. The Greeks and Phoenicians traded with, learned from and influenced each other, and the Greeks ended up taking the Phoenician alphabet and amending it to suit their own language. A clever Greek must have sat down one day and looked at the writing and thought, 'Yes, the sounds we make that we've never written down could be written down using this symbol code.'

So they systematically analysed each phoneme in Phoenician, along with the correspondence between these sounds and Phoenician letters; then did the same analysis with Greek, and eventually matched almost every phoneme in the Greek language with a Greek letter. They changed some of the letters for consonants which didn't exist in Greek, and used them instead for vowels. And so were born A (*alpha*), E (*epsilon*), O (*omicron*) and Y (*upsilon*). I (*iota*) was an innovation.

Immediately this new alphabet was also transferred to other places, notably Italy, where the Etruscans and then the Romans took it up and adapted it further. And the Romans gave it to us.

A final irony: the Greeks might never have used their alphabet, and so we might never have got ours because, at first, educated Greeks, most famously Socrates, considered their highly developed oral culture superior to a written culture. Socrates felt writing posed serious risks to society – he considered writing to be too inflexible, damaging to memory and he worried about superficial understanding by the untutored reader. Fortunately, Socrates' student Plato ignored his mentor and wrote all this down for posterity.

The Dead Sea Scrolls

In central Jerusalem there's an unearthly-looking white domed structure, a bit like a meringue spaceship. This is the Shrine of the Book, a special climate-controlled building where some of the world's most precious religious documents are housed. The most famous documents here are the Dead Sea Scrolls, 30,000 separate fragments making up 900 manuscripts of biblical texts and religious writings from the time of Jesus which were discovered in caves in the Judaean Desert. They're considered one of the most important archaeological finds of the modern era.

For the past eighteen years, the Israel Antiquities Authority has been working out how to preserve these priceless fragments of the past. They're so precious and so fragile that only four highly trained conservators are allowed to touch them with tweezers and rubber gloves.

The texts are of great religious and historical significance, as they include the oldest known surviving copies of biblical and extra-biblical documents. They're written in Hebrew, Greek and Aramaic,

A more complete fragment of the Dead Sea Scrolls

the ancient language we think Jesus spoke. Some texts are substantial and complete, the longest scroll being 8 metres long, but the majority have disintegrated and broken up over the centuries, and what remains are thousands and thousands of fragile pieces of a jigsaw. The texts were written over a period of around 200 years, up to about AD 70, the date of the First Jewish Revolt.

Carbon dating puts the earliest of them at about 150 BC. Scholars think they were placed in the caves to hide them from the advancing Roman army. The people who wrote them were clearly a sect who had had enough of Jerusalem and segregated themselves from mainstream Judaism. They wrote about their lives and ideas, and about their belief in a Messiah. It's these writings which are of particular interest to Christian scholars.

The scrolls contain all the books of the Hebrew Old Testament, except Esther and Nehemiah, many in several copies; religious texts which aren't part of the standard canon, such as Enoch and the Book of Jubilees; and other religious texts and secular writings including lists of laws, advice on warfare and a catalogue of places where treasure was buried.

How on earth did these ancient writings on parchment and papyrus survive? It was all about location. They were hidden in caves in the Judaean Desert, at Khirbet Qumran on the north-western coast of the Dead Sea, a low-lying area with a dry climate. The story of their discovery is a great yarn, no doubt embellished and added to over the years, like all the best stories. In 1947 a Bedouin shepherd looking for a lost goat came upon these caves. He threw a stone into one to see if the goat was there, and he heard a strange noise. The noise turned out to be his stone hitting one of the clay jars containing some scrolls. He went in to investigate and found a broken jar with scrolls inside. He took the scrolls and brought them to an antique dealer in Bethlehem, and the antique dealer, realizing what he had in hand, told him to go back and bring some more. So he went and brought more, and these are the first seven scrolls, parts of which are exhibited at all times at the Shrine of the Book.

The goatherd's name was Muhammad Ahmed al-Hamed, nicknamed Muhammad edh-Dhib. Another version of the story says that edh-Dhib's cousin, Jum'a Muhammad, saw some cave holes one day and threw a rock in to find out how big they were. He discovered they were big enough to fit a man and told his cousin what he had found. Edh-Dhib went and had a look, fell into one of the holes and retrieved some scrolls from a pot.

Once the initial discovery was made and pieces of ancient scrolls began to appear on the market and change hands, news of the finds spread, and gradually over the next few years dozens of caves were discovered, eleven of which contained pots of scrolls. The scrolls came to Israel by various routes. Four were sold through an advertisement in the *Wall Street Journal* in June 1954 and bought for the Israel government by the archaeologist Professor Yigael Yadin for US$250,000. They were brought to Jerusalem and put on display in the Rockefeller Museum.

Eventually, work on reconstructing and preserving the scrolls began. It was a hugely complicated, delicate and flawed process at the start, because no one knew the best way to do it. The first scholars at the Rockefeller Museum in Jerusalem spread the fragments on trestle tables, and every two pieces that they thought matched they Sellotaped. Sellotape was the latest invention then, and they had no idea it would be so destructive. It penetrates the parchment or the papyrus and causes its disintegration. Much of the work of preservation today is about undoing – literally – the damage done, because they're still trying to get the sticky mess off.

MISCELLANEOUS FOR SALE

"The Four Dead Sea Scrolls"

Biblical Manuscripts dating back to at least 200 BC, are for sale. This would be an ideal gift to an educational or religious institution by an individual or group. Box F 206, The Wall Street Journal.

The ad that appeared in the Wall Street Journal

Archaeologist Yigael Yadin scrutinizing a piece of scroll on which Hebrew writing is clearly visible

Sellotape wasn't the only well-meaning but harmful intervention. The British Museum sent a conservator in the 1960s to have a go with another treatment, which also proved to be quite damaging.

Different attempts to conserve and restore the scrolls continued through the 1970s and 1980s, and in the 1990s the project to publish them really got going. Scholars from all over the world were given access, provided they published, and so far, forty volumes of photographs of the scrolls, plus commentary, have been published by Oxford University Press.

As the team continued their work on repairing and preserving the scrolls, they were concerned in case they were, in their turn, causing more damage. They needed to know what was happening to the scrolls, what their underlying condition was, so they turned to state-of-the-art infra-red and multi-spectral multi-wavelength imaging, which can 'see' deep into the scroll and reveal things not visible in natural light – not just the physical condition of the parchment, but previously hidden words which allow scholars to interpret the texts afresh.

They went even further and, working with scientists and scholars, decided to digitize everything and make it available to the public online. The scrolls were photographed in the best-possible colour images, as well as infra-red, and published online alongside all of the translations, transcriptions, the bibliography and comments of the scholars. The imaging process shows up all sorts of things not visible before, including tiny details like the lines which the original copiers drew on the parchment before they started. Curiously, they wrote *below* the line, not on it, so it looks as if the words are suspended. All this – the scrolls in greater clarity than ever before, the transcription, the translation, the commentary – will be available online on your computer at home. Everyone, from the professional academic to the schoolteacher to the curious public, will have access. These precious, fragile bits of history, discovered by chance in the musty darkness of remote caves in the desert, have been preserved by the technology of the space age for the generations of the internet age, and beyond.

The scrolls are beautiful to look at, but they're very human, too, because they have mistakes in them. Parchment and papyrus were expensive and precious, so they couldn't just take a fresh sheet and start again. There are marks where someone has rubbed out his mistakes, and one where the word 'God' has been written in where it shouldn't have been, but because the word was 'God', the scribe couldn't just rub it out, so instead he marked it so that the person reading it out loud would know not to say it here. Literally, the word of God.

Stephen Fry examines a scroll at the Israel Antiquities Authority

Pinyin

Sir David Tang sits in the Luk Yu Tea House in Hong Kong. It is a throwback to colonial times – not the haunt of the *guilo* foreign devil (the white man) but of the old Hong Kong Chinese. Sir David is one of the most colourful and intelligent of its denizens and a true cosmopolitan. Born into a wealthy Hong Kong Chinese family, he was shipped off aged thirteen to boarding school in Cambridge and speaks perfect English. But he also speaks Cantonese, his mother tongue, which is spoken by some 70 million Chinese, and Mandarin, which is the official dialect of the People's Republic of China (PRC), spoken by some billion people. He's joined by his old chum Johnson Chang, and together they proceed to throw light upon the complexity of the Chinese language.

But there's a problem with even saying this, as they explain that Chinese is not one language but a language family of some thirteen sublanguages, which are mostly mutually unintelligible. So a Cantonese-speaking person in Hong Kong cannot understand someone from Beijing, Shanghai or Chengdu. However, the good news is that it's all written pretty much the same way, so although it sounds very different it can be read by everyone. Since Mao's revolution in 1949, Mandarin has become the standardized form of spoken Chinese and is based on the Beijing dialect – not a popular move with the Cantonese, according to Sir David.

Johnson is a skilled calligrapher and demonstrates the rudiments of the writing system. Chinese characters, *hanzi*, are written within imaginary rectangular blocks, traditionally arranged in vertical columns, read from top to bottom down a column, and right to left across columns. Chinese characters are morphemes, units of language which are independent of phonetic change. As he dips his brush and with a few elegant strokes makes ever-more complex characters, Johnson explains that Chinese characters represent the oldest continuously used system of writing in the world, which has remained virtually unchanged for 3,000 years. The number of Chinese characters is approximately 47,000, although a large number of these are rarely used variants accumulated throughout history. Fortunately, literacy requires a knowledge of only between 3,000 and 4,000 characters. The characters are *morphosyllabic*, each usually corresponding to a spoken syllable with a basic meaning. However, although Chinese words may be formed by characters with basic meanings, a majority require two or more characters to write and have a meaning that is distinct from the characters they are made from. With a few more brush strokes Johnson illustrates how the earliest characters were created as pictograms, similar to hieroglyphs, but explains that, contrary to popular belief, pictograms make up only a very small portion of Chinese characters. While characters in this class derive from pictures, they have been standardized, simplified and stylized to make them easier to write, and their derivation

Chinese oracle bones from the Shang Dynasty

is therefore not always obvious. Most characters contain phonetic parts and are composites of phonetic components and a basic root or radical. Only the simplest characters, such as *ren* <人> (human), *ri* <日> (sun), *shan* <山> (mountain), *shui* <水> (water), may be wholly pictorial in origin. Only around 4 per cent are pictographs, and 80–90 per cent are phonetic groupings consisting of a semantic radical element that indicates meaning and a phonetic element that indicates the pronunciation. There are about 214 radicals, and the rest of the characters are built upon them.

So the ideograms either modify existing pictographs or are direct symbolic illustrations. For instance, by modifying <刀> *dāo*, a pictogram for 'knife', by marking the blade, you get an ideogram <刃> *rèn* for 'blade'. The ideogrammic compounds symbolically combine pictograms or ideograms to create a third character. For instance, doubling the pictogram <木> *mù*, 'tree', produces <林> *lín*, 'grove', while tripling it produces <森> *sēn*, 'forest'. Similarly, combining <日> *rì*, 'sun', and <月> *yuè*, 'moon', the two natural sources of light, makes <明> *míng*, 'bright'.

Zhou Youdong,
creator of Pinyin

Not surprisingly, literacy rates were very low when Mao came to power. Sir David lets on that the man who more than anyone changed the way Chinese now learn to read and write is still alive. His name is Zhou Youdong, and he was entrusted by Mao with the total overhaul of the Romanization of Chinese.

Zhou Youdong is 106 years old but looks a decade or three younger. He has survived more revolutions, counter-revolutions, great leaps forward and giant leaps backwards than Newcastle United FC. He lives with his eighty-five-year-old son in one of those drab communist-era housing projects that so disfigure the Chinese capital.

Zhou was living in the USA before Mao came to power, studying economics. But in 1950 he decided to return as he was keen to help in the modernization programme of the communists. At that time there was a rather complicated system of Romanization of the written language which was not very effective. The difficulty of learning the characters had held back literacy rates so that when Mao came to power only 20 per cent of the country was literate. It was very important to improve this if China was to become a modern state. In 1955 Mao asked Zhou to make a new system; it took him and his colleagues six years to complete. Two decades after the revolution literacy rates had improved to 80 per cent and now they are about 90 per cent. Without doubt Pinyin, as Zhou's Romanization is called, has done more for developing China than almost anything else one can think of. Yet Zhou remains modest about his achievements. Was Mao Tse Tung grateful? Diplomatically Zhou answers, 'Mao Tse Tung supported the work.' Pinyin is now taught in every school in China, and its phonetic-based Romanization system allows children to write Mandarin (on which it is based) much faster than the old way of learning all the characters by heart. Moreover, it revolutionized the way you type, especially important in a country that texts more than the rest of the world combined. Without Pinyin this would simply not be possible.

There had been an earlier Anglicization of Chinese made by Thomas Wade in 1859 and modified by Herbert Giles in 1892, but it was more for foreigners than for Chinese. This system uses English consonants and vowels to approximate the phonology of Mandarin

Chinese. This is the system that gives us *Peking* and *Mao Tse-Tung* rather than *Beijing* and *Mao Zedong*. In truth, neither of these systems really reproduce the sound of the Chinese, so we might as well stick with saying *Peking* rather than mangle the Mandarin of *Beijing*, which will sound all wrong, as we can never get the tone right.

Printing

We've seen how the invention of writing changed the human world utterly. With writing, we could preserve, record and shape our stories, communicate our ideas and extend the significance of our lives beyond their end. Writing gave us our history.

Over the 5,000 years since writing was invented, hundreds of different languages have been written in hundreds of different scripts. For the vast majority of this time, everything was written by hand. But nearly 600 years ago there was a revolution in writing technology which changed, not just the face of writing, but the whole world. In the mid fifteenth century a man called Johannes Gutenberg invented a mechanical way of making books. His development of movable type printing started the printing revolution and is widely regarded as the most important event of the modern period.

Gutenberg's achievement wasn't that he was the single genius who invented printing in one inspired moment. Forms of printing had been used in different parts of the world long before Gutenberg came along. The Chinese had been working with type carved into wood and bronze for centuries. They had also invented paper made from a pulp of water and discarded rags which was then pressed into sheets for writing or printing on.

In the British Museum there's a copy of the *Diamond Sutra*, one of the most sacred Buddhist texts and the world's earliest complete survival of a dated printed book. The original was made in 868 and hidden for centuries in a sealed-up cave in north-west China. It's made from seven strips of yellow-stained paper which were printed from carved wooden blocks and pasted together to form a scroll over 5 metres long.

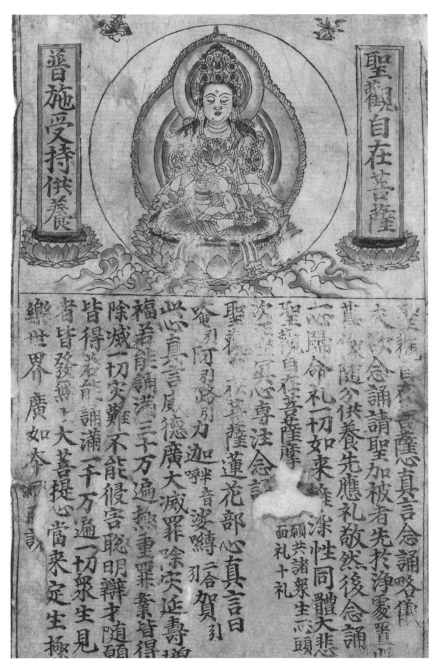

Diamond Sutra, *the oldest surviving printed book*

279

Later on, the Chinese developed a movable type system using porcelain characters, and in the thirteenth century the Koreans were already experimenting with metal movable type made out of copper. The screw press, which Gutenberg refined and developed, had also been around for centuries to press grapes and impress patterns on textile.

What Gutenberg did was to combine different crucial elements and adapt them. He invented a process for mass-producing movable type – the individual pieces of type in metal, one for each character of the alphabet, punctuation and other signs, which could be set up to be printed on a printing press, and then reused over and over again. He used new oil-based instead of water-based ink. And he had the radical idea of using the traditional agricultural screw press for printing; up to then, impressions had been made from stamps or wood blocks, by pressing them on to paper or cloth or by putting paper on top of them and then rubbing to get an impression.

He brought these techniques together into a practical system which, as it developed, became increasingly efficient, economically viable and fast. In doing so he laid the foundation for mass production of books in Europe.

His printing press, which was in operation by 1450, played a key role in the development of the Renaissance, Reformation and the scientific revolution because it made possible the spread of learning and the sharing of knowledge on a huge scale. The written word escaped from the quill pens of the monks in their scriptorium and took flight out into the wide world.

And yet we don't really know very much about Gutenberg himself. He was born in the German town of Mainz in 1400 and died, a poor man, sixty-eight years later. In between, he sort of flitted about, disappearing for periods, re-emerging somewhere else, always working on something, always short of money, frequently involved in some dispute or quarrel. At one stage he was sued for debt by the man who financed his printing press, and he ended up losing the press in lieu of payment. We don't know whether he was married, or whether he had any children, but his achievements were recognized at the end of his life when he was given the title *Hofmann* (gentleman of the court), an annual stipend, 2,180 litres of grain and 2,000 litres of wine tax-free. He was buried in the Franciscan church at Mainz, but the church and the cemetery were later destroyed, and Gutenberg's grave is now lost.

The First Books

The first real book to be printed using the new techniques Gutenberg developed in the 1450s was the Gutenberg Bible, also known as the forty-two-line Bible. No one knows exactly how many copies he printed, but the best guess is around 180 – 145 on paper and the rest on the more luxurious and expensive vellum. Only forty-eight copies survive, not all of them complete, and scholars consider them to be among the most valuable books in the world. The Bible is a thing of beauty, printed in Vulgate Latin on hand-made, fine-quality paper from Italy in Blackletter font on a page design that was deliberately created to look like an illustrated manuscript. Each copy was sold in folded sheets and was later bound and decorated according to the wishes of its owner. They cost much less than a handwritten Bible, but at least some copies are known to have sold for 30 florins, a price beyond most students, priests or other people of ordinary income. Probably most of them were sold to monasteries, universities and particularly wealthy individuals.

The price of books began to fall as the technology of printing improved and the number of printing presses all over Europe increased. But although the process of printing was new, in the early years the printer was in direct competition with the scribe and he wanted to make books, like the Gutenberg bible, that were as luxurious and beautiful as the handwritten ones.

So printers left large areas of the page blank so that an illuminator could decorate the text afterwards and create the same ornate, elaborate capital letters and richly illuminated patterns as the manuscripts. The same Gothic characters were reproduced in the new fonts, and the printer would even join up some of the letters so that they looked as if they'd been written by a quill pen. It was as if the printer was saying: 'Look, this is new and it's the future – but it's

Gutenberg's Bible with hand-illuminated letters to help ease the transition to print

282

just as splendid and well produced as what you're used to.'

The nineteenth-century writer, historian and social commentator Thomas Carlyle is pretty good for a quote on most things, and his comment in *Sartor Resartus* (1833) on the invention of the printing press has all the drama and flair you'd expect: 'He who first shortened the labor of copyists by device of movable types was disbanding hired armies, and cashiering most kings and senates, and creating a whole new democratic world: he had invented the art of printing.'

It's the creation of a new 'democratic world' of the free circulation of knowledge and learning that's at the heart of the printing revolution. As printing spread, it created a wider literate reading public, and it vastly increased the range of books and subject matter. The Church had dominated book production up to then, authorizing what was to be copied and controlling access to learning. But now, more secular books began to be printed, and religious books, bibles (like the pocket versions by the Venetian printer Aldus Manutius) and tracts could be printed in great numbers and widely disseminated. Now people could read and decide for themselves.

This was crucial to the success of the Reformation in Europe. Martin Luther was armed not just with radical ideas, but with a printing press. It's easy to burn heretical texts which have a few hundred manuscripts in circulation; but what does the Church do when thousands of texts can be sprayed around quickly and at a fraction of the cost? William Tyndale, the scholar and translator, found out when he challenged the authority of the Roman Catholic Church and the English Church and state by translating the Bible into English. Copies of his Bible were secretly published and passed around Europe, and he was tried for heresy and burned at the stake in 1536. The French humanist scholar and printer Etienne Dolet was twice condemned to death and twice pardoned for his printing of classical works and authors like Rabelais and Calvin. He was eventually burned at the stake for heresy in Paris in 1546. He's often called the first martyr of the Renaissance. Printing spread learning and enabled the radicalism of the Reformation and the humanist ideas of the Renaissance to percolate through all levels of society. In Britain it also had a profound effect on the written language.

Geoffrey Chaucer was the first English writer to be set in print, though, as he died just around the time that Gutenberg was born, he missed the print revolution. But he certainly made it clear that he was pretty fed up with the sloppiness of the scribes who copied out his works for readers. In fact, in one of his great poems 'Troilus and Criseyde', there's a bit at the end when he writes a little verse to the reader which shows his anxiety about the lack of standardization in written and spoken English. He basically asks that his poem isn't too badly mangled by the copyist: 'for there is so great diversity in English and in writing of our tongue so pray I God that none miswrite thee little book. Neither miss meter for default of tongue and where so thou be or else sung that that thou be understonged. God I beseech.' In other words he hoped that people would find some way of spelling all the different words at least in such a manner that it was generally understood by those who were going to read it.

And that's precisely what printing allowed. Chaucer's *Canterbury Tales* was the first book by an English author that the printer William Caxton published in England. Caxton started printing at a time when the English language was going through a period of massive upheaval. It was changing from the Middle English of Chaucer to the Modern English we speak today – a process that's called the Great Vowel Shift. The 'continental' pronunciation of sounds based on Latin and Italian gave way to more regional lilts at about the same time written English was catching on through the printing press.

This was a time when the English language was incredibly diverse. Dialects of different regions had different words for the same thing, and as for spellings of the same words, well, there were over twenty different spellings just for the word *might*! In order to sell as many books as possible, Caxton needed a linguistic norm for the vernacular. He needed what Chaucer was looking for in his complaint at the end of 'Troilus and Crisedye' – a standardized, consistent spelling that could be understood by everyone. So Caxton printed the dialect of London and the South-east Midlands (though he dumped his Kentish *eyren* in favour of northern *egges*). Through the medium of print, Caxton made people across the country familiar with

Of eche of them so as it semed me
And whyche they were and of what degre
And in what aray eke they weren ynne
And at a knyght thenne I wyl begynne

A Knyght ther was a worthy man
That fro the tyme that he first began
To ryden out / he loued chyualrye
Trouthe & honour fredom and curtesye
Ful worthy he was in hys lordis werre
And therto hadde he ryden noman ferre
And as wel in crystendom as in hethenesse
And euer honoured for hys worthynesse
At alisaundre he was whan it was wonne
Ful ofte tyme he hadde the boord begonne
Abouen alle nacions in pruce
In lettowe hadde he reysed and in Ruse

Prologue to Chaucer's Canterbury Tales, *1485*

I rather than *ich* and *home* rather than *hame*. On the one hand he preserved old spellings that no longer matched pronunciations; on the other hand he started to modernize prose. Printing added an element of linguistic stability to literature. No doubt Chaucer would have approved.

Diderot and the *Encyclopédie*

Printing didn't just give rise to greater literacy; it was one of those inventions, like gunpowder and the compass, which changed the state and appearance of the world. With more books available to more people, reading went from being an exercise of the elite, to a far more accessible activity. And because more people could read more, it changed attitudes towards learning and knowledge. As printing spread, the costs decreased, and booksellers proliferated, especially in the big European capitals, like Paris and London. The printed word became the life blood of the age of reason and inquiry we call the Enlightenment.

The booksellers of Paris have long been part of a kind of literary underworld, spreading subversive ideas by printed pamphlets, books, leaflets and newspapers. And in the eighteenth century the cafés, or coffee houses, were the places where all these ideas were discussed. In Paris and in London like-minded thinkers congregated to read, as well as learn from and to debate with each other. The Café Procope is one of the oldest and most famous, where writers, philosophers, intellectuals, politicians and anyone who liked an argument would hold court and exchange ideas and gossip. 'There is an ebb and flow of all conditions of men, nobles and cooks, wits and sots, pell mell, all chattering in full chorus to their heart's content,' was how one contemporary writer described the scene. Voltaire used to go there and, so the story goes, get through forty cups of coffee a day, mixed with chocolate. Thomas Jefferson, Benjamin Franklin, John Paul Jones, Jean-Jacques Rousseau – they were all players at the Procope.

This was café society at its most potent, a sort of borderless intellectual republic of letters and ideas and free thought where the driving forces of the Enlightenment – the critical questioning of traditional institutions, customs and morals, the central belief in rationality and science – had free rein. In a way, this period began the process that has ended with the internet and social networks.

The Café Procope in Paris, meeting place for revolutionary thinkers

What better place, then, to find Dr Kate Tunstall, an Enlightenment scholar, to talk about the book that embodied the new way of thinking and the extraordinary man behind it – the *Encyclopédie* and its editor, Denis Diderot. The *Encyclopédie – ou Dictionnaire Raisonné des sciences, des arts and des métiers* ('Encyclopaedia, or a Systematic Dictionary of the Sciences, Arts and Crafts') was originally commissioned as a translation into French of the *Cyclopaedia* of Ephraim Chambers, which first appeared in 1728, but in the hands of the polymath Denis Diderot it grew into something huge, radical and utterly contemporary.

'Diderot knew English and he was also known to be subversive and, therefore, marketable,' says Kate. The booksellers behind the project were businessmen as well as publishers, and they had an eye to a successful commercial enterprise. And if they were looking to make a splash with their main choice of editor, Diderot was the right man.

Denis Diderot was a novelist, playwright, art critic, man of science and philosopher – and the most talkative man of his generation, according to one contemporary. The Enlightenment was a time when science and reason began to challenge the authority of the Church, and Diderot worked under threat of exile during most of his life because of his controversial, subversive views. In fact, he landed in prison for three months not long before the first volume of the *Encyclopédie* came out in 1751. 'We know he had these dinners at his home where he openly declared himself an atheist,' says Kate, 'and one of the reasons he was arrested, we think, is that his parish priest had been reporting on him to the police.'

The *Encyclopédie* publishers were extremely upset at Diderot's imprisonment because they feared financial ruin. 'So they wrote to the authorities,' continues Kate, 'saying, "Please let him out." And one of the things they explain is that, if they had to bring all the manuscripts to Diderot in prison, even at this early stage, they would fill a whole room! So you have to imagine Diderot sort of surrounded by paper, trying to fit it all together.'

It's a vivid image of just how massive an undertaking the *Encyclopédie* was. Diderot, and his co-editor d'Alembert (who left the project after a couple of years), commissioned essays and articles, not just on the academic and scientific knowledge of the day, but on the technologies, industries and trades of the working people. 'An encyclopedia . . . should encompass not only the fields already covered by the academies, but each and every branch of human knowledge,' Diderot wrote. And by doing so, he maintained, it would 'change the way people think'.

And it would do it, too, by influencing opinion and thought, by challenging authority, introducing new ideas, looking at things differently. Diderot was clever, though, in *suggesting* radical ideas rather than headlining them. He used irony and strategies of

ENCYCLOPÉDIE,

OU

DICTIONNAIRE RAISONNÉ

DES SCIENCES,

DES ARTS ET DES MÉTIERS,

PAR UNE SOCIETÉ DE GENS DE LETTRES.

Mis en ordre & publié par M. *DIDEROT*, de l'Académie Royale des Sciences & des Belles-Lettres de Pruſſe; & quant à la PARTIE MATHÉMATIQUE, par M. *D'ALEMBERT*, de l'Académie Royale des Sciences de Paris, de celle de Pruſſe, & de la Société Royale de Londres.

Tantùm ſeries juncturaque pollet,
Tantùm de medio ſumptis accedit honoris ! HORAT.

TOME PREMIER.

A PARIS,

Chez
{ BRIASSON, *rue Saint Jacques, à la Science.*
DAVID l'aîné, *rue Saint Jacques, à la Plume d'or.*
LE BRETON, Imprimeur ordinaire du Roy, *rue de la Harpe.*
DURAND, *rue Saint Jacques, à Saint Landry, & au Griffon.*

M. DCC. LI.

AVEC APPROBATION ET PRIVILEGE DU ROY.

Denis Diderot's Encyclopédie, *1751*

subversion within the text. Articles say that they're going to be about one thing, and then they turn out to be about something slightly different. Or they have cross-references to a directly opposing point of view. That was Diderot's way of changing the normal way of thinking. So if you look up 'cannibal', for example, there's a cross-reference to Eucharist and Communion – so he's making a point about, or mocking the idea of, the eating of the body and blood of Christ in church. And there's a streak of mischievous wit, too, in some of the entries where Diderot plays about with the notion of what should and shouldn't be in an encyclopaedia, and what some people are expecting to find. Kate's favourite one is 'aguaxima': some kind of plant found in Brazil. But Diderot's entry (and we know it's his because it's preceded by an asterisk, which was his sign) is completely tongue-in-cheek. He has nothing interesting to tell about it, so he just mocks the convention of including it in an encyclopedia at all. This is plainly not the challenging and ground-breaking material he was interested in.

Aguaxima, a plant growing in Brazil and on the islands of South America. This is all that we are told about it; and I would like to know for whom such descriptions are made. It cannot be for the natives of the countries concerned, who are likely to know more about the aguaxima than is contained in this description, and who do not need to learn that the aguaxima grows in their country. It is as if you said to a Frenchman that the pear tree is a tree that grows in France, in Germany, etc. It is not meant for us either, for what do we care that there is a tree in Brazil named aguaxima, if all we know about it is its name? What is the point of giving the name? It leaves the ignorant just as they were and teaches the rest of us nothing. If all the same I mention this plant here, along with several others that are described just as poorly, then it is out of consideration for certain readers who prefer to find nothing in a dictionary article or even to find something stupid than to find no article at all.

Diderot was the spider in the middle of this vast web of ideas, essays, learning and information, receiving articles from about 150 contributors, writing many himself, editing, arranging and cross-referencing them. There was a huge range of contributors, from big names like Voltaire and Rousseau to doctors, chemists and academics we don't know much about and who are not named. Rousseau wrote some of the articles about music, but he went off in a huff after d'Alembert wrote an entry about Geneva, Rousseau's home town, and said it was a very boring place that needed a theatre. The *Encyclopédie* didn't just include written descriptions of the sciences, arts and crafts, but hundreds of plates that illustrate the subjects. Diderot got into trouble for being rather vague about the sources of some of the illustrations and diagrams, and he was accused of plagiarism, but the plates are a magnificent feature.

Diderot was particularly fascinated by what was called 'the mechanical arts', all the trades and artisan crafts of the time. He was surely influenced by his own background. His father, Didier, was a master cutler, a maker of cutting tools, in the provincial French town of Langres, and was particularly well known for surgical knives, scalpels and lancets, which were stamped with his hallmark of a pearl. The young Diderot must have visited his father's workshop and watched the cutters at their work, grinding the knives and forging blades, and he himself wrote the article about 'cutler'.

There are dozens of other plates, illustrating everything from tile-making to how pack saddles and harnesses are assembled; from the work of the button-maker and confectioner to cabinet-making and the silk factory. There are plates showing tools, diagrams of machinery, people at work and factories, and the fact is, no one had really seen these things illustrated before. The tradition up to then had been to illustrate the *beaux arts* and nature – not the craft of the working man and woman.

The *Encyclopédie* was an astonishing achievement. It was only supposed to be a few volumes at the start; in the end, Diderot devoted twenty-five years of his life to it and produced thirty-five volumes, with 71,818 articles and 3,129 illustrations between 1751 and 1772. It is a work which stands for the Enlightenment in its scope and curiosity, its boldness and its thirst for knowledge and the free exchange of ideas. It's big and sprawling, a baggy monster – and you can look at it for yourself online.

The nineteenth-century literary critic C. A. Sainte-Beuve wrote of the *Encyclopédie*: 'It has been compared to the impious Babel; I see in it rather one of those towers of war, one of those siege-machines, enormous, gigantic, wonderful to behold . . .' In a way, the *encyclopédistes* were laying siege to their world of the eighteenth century, looking it full in the face, surveying it, undermining it and challenging accepted wisdom. Truly capturing their world in words.

Dr Johnson's Dictionary

In the middle of the eighteenth century, while Denis Diderot was assembling his vast *Encyclopédie* just across the Channel, an English man of letters was single-handedly writing another reference book, so famous that it's simply referred to as 'Dr Johnson's Dictionary'.

Samuel Johnson was one of the towering figures of the Enlightenment, a man of enormous wit, learning and literary brilliance. Many TV viewers find it difficult to think of him without remembering that episode of *Blackadder* with Johnson, played by Robbie Coltrane, as an annoying show-off obsessed with his great work; he comes to tell Hugh Laurie's Prince George all about his Dictionary – and Baldrick uses the manuscript to light the fire. The real Dr Johnson was, doubtless, a dinner party bore about his Dictionary but with good cause. No one had ever written a dictionary like it before. There had been several attempts before *Dictionary of the English Language* was published in 1755, but they were really just markers along the way. A schoolmaster called Robert Cawdrey had written a book of word definitions in 1604, but it only had a couple of thousand or so entries in it. Twenty years later, another scholar, Henry Cockeram, had a go at improving things and was the first to call his book a 'Dictionary'. Others made their contributions, too, but by the middle of the century there was still no authoritative lexicon of the English language. By contrast, the French and Italians were cantering ahead with academies and armies of scholars dedicated to the production of detailed, accurate dictionaries.

Britain needed a proper dictionary; it was a matter of national pride and academic necessity. It was also a tremendous commercial opportunity in a society where books and reading were the life blood of the Enlightenment. So it's no surprise that the impetus came from a group of booksellers – who were also publishers in those days – with an eye on the market.

The consortium was led by one of the foremost publishers of the day, Robert Dodsley. He had been a playwright before he went into bookselling, and published all the big names, including Alexander Pope, Thomas Gray and Oliver Goldsmith. He also knew and had published Johnson, and he picked him as the ideal man to take on the dictionary. Johnson was a prolific writer, at this time in his early career (he was only thirty-seven) a hack, someone who earned his living and put food on the table entirely through the power of his pen. It's said he could churn out 10,000 words a day on anything and everything, in the form of magazine articles, parliamentary sketches, the odd poem and verse drama, prefaces, introductions. For two years he single-handedly wrote an entire, twice-weekly periodical, *The Rambler*. His industry and output were prodigious, and he had a reputation for a thorough and committed approach to every writing commission.

Dodsley met with Johnson, and the deal was struck: 1,500 guineas to produce *the* English dictionary. It was a handsome fee which allowed Johnson to set himself up in a house off Fleet Street and dedicate himself – for the next nine years – to his monumental work.

How on earth does one man set about producing a dictionary of the English language? When he signed the contract, Johnson seemed undaunted. He declared he would complete it in three years and mocked the French Academy for appointing forty scholars who took forty years to write their dictionary. 'This is the proportion,' he commented drily. 'Let me see; forty times forty is sixteen hundred. As three to sixteen hundred, so is the proportion of an Englishman to a Frenchman.'

In fact, it took Johnson three times as long because he quickly realized that his initial plan was flawed. At first he began trawling through other dictionaries, starting with 'A' and plucking out lists of words, but almost immediately he decided he had got things the wrong way round. Instead, he started from words in *action*: he turned to books, to English literature, and selected his words from those sources. In this way he put together his wordlist from illustrations, rather than listing words and then finding examples. Workhorse that he was, he tackled the challenge head on: he simply immersed himself in reading and books for years and years. Everything from medical tracts, plays and novels to political pamphlets, poetry and theology written over the previous century and a half, a staggering 2,000 books, nibbled, tasted, chewed or swallowed whole in his search for examples of how we used our language.

Johnson ranged widely across topic and author, but the vast majority of the 100,000-plus quotations he used were drawn from the big guns

Dr Samuel Johnson assembled the first major English-language dictionary

of literature: the Bible, Shakespeare, John Milton, John Dryden, Jonathan Swift. And from these illustrations he assembled a dictionary of 42,773 words, many of them with multiple entries to cover the range of meanings of the same word.

That was another discovery he made early on in his work: he was not content to produce a dictionary that was prescriptive, i.e. which set out how the language *should* be. He wanted it to be descriptive – to be faithful to the richness and flexibility of English as it was used. When he was drawing up his plan at the start, he thought that a word might have, at the most, seven or eight different meanings: in the event, he found words with twice or even three times as many subtle differences of usage. For the word *put*, he included 134 different meanings, among them to 'lay' or 'place'; to 'repose'; to 'urge'; to 'state'; to 'unite'; to 'propose'; to 'form'; to 'regulate' – the list goes on and on. *Time* had twenty definitions and fourteen illustrative quotations; *turn* had sixteen definitions and fifteen examples.

The point was, here was language sorted and illustrated by the standards of literature – and for literature read 'how the language was actually being used, in all its invention and shades and evolutions of meaning' – not language as it ought to be, or language bound for ever to some elegant, academic fixed standard. In his preface to the book Johnson described the 'energetic' unruliness of the English tongue. 'Wherever I turned my view,' he wrote, 'there was perplexity to be disentangled, and confusion to be regulated.' His original aim, he said, was 'to refine our language to grammatical purity'. But it was his genius to recognize that language is impossible to set in stone, because its nature is to change and evolve. And so, he said, he realized his role was to *record* the language of the day, rather than to form it. It was an approach which would have a lasting influence on the future writing of dictionaries.

Johnson didn't work entirely alone. He employed a handful of secretaries and assistants – all but one hailed originally from Scotland – to collate the thousands of passages and words he marked up in his reading. He turned the top floor of his house into a study and his atticful of Scotsmen laboured through the mind-numbing process of cutting up the definitions and quotations into little slips, ordering and arranging them, copying them into notebooks and adapting the process as the dictionary grew ever larger. Imagine the energy and dedication, the sheer concentration required to continue this pains-taking, repetitive, exhausting work day after day for nine years. No wonder Johnson was thoroughly fed up – even depressed, prob-ably – by the end. One of the illustrations he used for the word 'dull' he made up himself: 'To make dictionaries is dull work', and another famous one is his definition of a 'lexicographer': 'a harm-less drudge'. And in the preface he doesn't pretend that the process has been anything other than long, hard and draining: 'I have protracted my work,' he writes, 'till most of those whom I wished to please, have sunk into the grave, and success and miscarriage are empty sounds.'

However hard it was to compile, it's a brilliant, endlessly fasci-nating book to dip into. There's such flair and wit, such colour and life in it. There is an ease and facility in Johnson's writing, an economical elegance of some of his definitions. Take *dotard*:'one whose age has impaired his intellect'; *embryo*: 'the offspring yet unfinished in the womb'; *envy*: 'to repine at the happiness of others', or the words which show an earlier, more literal meaning, like *eavesdropper*: 'a listener under windows'; or *jogger*: 'one who moves heavily and dully'. And then there are the curmudgeonly, witty definitions which are one part linguistic definition and three parts personal prejudice. The pleasure is not in the accuracy or strict factual truth of the definition, but in the amusement and entertainment it provides, like the famous *oats*: 'a grain, which in England is generally given to horses, but in Scotland supports the people'; or *excise*: 'a hateful tax levied upon commodities and adjudged not by the common judges of property but wretches hired by those to whom excise is paid'.

When the Dictionary was published in 1755, two volumes' worth and a total of 2,300 pages, it was generally well received. Johnson was certainly criticized for various omissions and errors, like there being no entry for a simple word like *irritable*, and *leeward* and *windward* being given the same definition. The etymology was sometimes wayward – the historian Thomas Macaulay called John-son 'a wretched etymologist' – and one philologist declared that the grammatical and historical parts of the Dictionary were 'most truly contemptible performances'. And, of course, what makes the Dictionary fascinating and delightful and creative – the fact that it's the vision and execution of one man – is exactly what makes it flawed.

But the magnitude and breadth and flair of the Dictionary are undeniable. It is an astonishing work of individual scholarship and genius, the influence of which has stretched across the world and even further across the centuries. Noah Webster, the editor of the American *Webster's Dictionary*, disliked many aspects of Johnson's Dictionary, but he, too, liberally helped himself to verbatim chunks of it for his new work.

Perhaps the most curious footprint left by Johnson's Dictionary is to be found, of all places, in the US Constitution. The Dictionary's biographer, Henry Hitchings, tells us that American lawyers still refer to Johnson in questions of interpretation because, when the Constitution was originally drawn up, in the late 1780s, it was his Dictionary which was the prevailing authority on the English language. He cites an example from 2000, when American lawyers were debating whether US airstrikes against Yugoslavia violated the Constitutional principle that Congress alone can declare war. One of the central issues was: what *was* the authentic eighteenth-century meaning of *declare*: if you declared war, did it mean actual military engagement, or was it simply an acknowledgement of the necessary prior conditions for conflict? They turned to Johnson for guidance as to what the words of the Constitution meant. That would no doubt have given Dr Johnson a good deal of satisfaction.

Not Just a Dictionary but a National Institution

The story goes that the first editor of the Oxford English Dictionary kept a copy of Samuel Johnson's Dictionary open on his desk as he worked, marking all his definitions he included in the new tome with a 'J'.

The *OED*, as it's more commonly known, is considered to be *the* authority on the English language. Quite simply, it's one of the wonders of the world of words. Its second edition, printed in 1989, is the definitive guide to the meaning, history and pronunciation of over half a million words, past and present. These words make up 21,730 pages, bound in twenty dark Oxford-blue volumes, and take up four feet of shelf space. And the shelves themselves need to be pretty robust, given that the *OED* weighs almost 140 pounds.

The *OED*'s policy is to record a word's most-known usages and variants in all varieties of English. It's a constantly evolving process, as new words enter our language all the time – and for this, the online version is ideal.

The *OED*'s chief editor, John Simpson, was in charge of bringing the entire second edition of the dictionary into the internet age in 2000. John is now managing the first complete revision of the twenty-volume dictionary since it was originally published in 1933. His team started work at the letter M and they release updated sections of words every three months; in March 2011 they'd reached *Ryvita*. Computers and the internet may make revision physically easier and quicker, but up-to-the-minute access to rapidly expanding historical databases – from eighteenth-century farming manuals to twenty-first-century rap lyrics – means this is still a massive archaeological dig into the English language.

John Simpson first started working on the *OED* in 1976, in the pre-computer age; back then, the process of compiling the dictionary with the aid of millions of index cards hadn't really changed since the nineteenth century, when it all began. Below his office is the archive room where some of the index cards from the first dictionary have

James Murray, the Oxford English Dictionary's *first editor, surrounded by boxes of definitions*

been preserved: 700 boxes filled with neatly tied packets of diction-ary quotation slips and a topslip with the definition – many of them in the tiny handwriting of the *OED*'s extraordinary first editor, James Murray.

This white-bearded lexicographer gazes from photographs on the wall, a dead ringer for Professor Dumbledore. Murray was the Scottish schoolteacher and self-taught philologist (he claimed a working knowledge of twenty-five languages including Russian and Tongan) who was invited by the Philological Society of London in the late 1870s to realize their vision of a new dictionary to replace the existing outdated and incomplete dictionaries of Samuel Johnson and others.

The Oxford University Press was to publish it in instalments, and whereas Johnson's dictionary had drawn heavily on a main group of literary sources – Shakespeare, Dryden, Milton, Addison, Bacon, Pope and the Bible – the new collection of words and meanings would be supported by evidence drawn from a myriad of works and sources. Murray thought that the task could be completed in ten years and at first planned to keep on his job as a teacher at Mill Hill School whilst editing the dictionary at home. Little did he realize the gargantuan task he faced, both in collecting the quotations and then filing and editing them.

In April 1879, a month after his appointment, an appeal went out 'to the English speaking and English reading public' for a thousand people to 'read books and make extracts for The Philological Soci-ety's New English Dictionary'. A list of books, starting with Caxton's printed works, was included. All offers to help were to be addressed to 'Dr Murray, Mill Hill, Middlesex, N.W.'

In preparation for the flood of responses, and to store the sacks of quotation slips already amassed, Murray built a corrugated iron shed in his front garden, jokingly called the scriptorium. After six years, when the dictionary had only reached *A for Ant*, Murray gave up his teaching job and moved his family to a bigger house in Oxford, building a larger version of the scriptorium in his back garden.

It's here that the photographs of Murray, the white-bearded sorcerer of words, were taken. Perched on his head is the black

James Murray and his team of assistants in his specially built shed

cap in the style of his hero, the Protestant reformer John Knox, which he insisted on wearing to work every day. Surrounding him are shelves and pigeonholes bulging with millions of quotation slips illustrating the use of words to be defined in the dictionary. At one point these slips were arriving from volunteer readers at a rate of 1,000 a day; Murray even roped his eleven children into the work of sifting and alphabetizing them alongside his small team of assistants.

One of Murray's most prolific early contributors was an American surgeon by the name of Dr William Chester Minor. Over the years he and Murray corresponded regularly. Minor gave his address as Broadmoor, Crowthorne, Berkshire, so Murray must have known that he was connected to the Broadmoor Hospital for the Criminally Insane. But it wasn't until 1891, when Murray visited Dr Minor, that he learned the full story. Minor, a surgeon in the Union army during the American Civil War, had been committed to Broadmoor in 1872 after shooting dead a stranger, George Merrett, in London.

Minor lived a fairly comfortable life at Broadmoor. He had two rooms – a bedroom and a dayroom – and a private income which allowed him to buy books and good food. He sent money to the widow of the murdered Merrett, and it's said that the two got on so well that she visited him monthly, bringing him new supplies of books. In 1879, Minor heard about the appeal for readers for the new English dictionary and he set about the task which dominated most of the remainder of his life. Over the next twenty years, Minor supplied tens of thousands of quotations. James Murray described his contributions to the writing of the dictionary as 'enormous', acknowledging that, in a two-year period, Minor had sent in at least 12,000 quotations.

A typical submission from Minor to James Murray is his contribution to the verb *to set*: 'a1548 Hall Chron., Hen. IV. (1550) 32b, Duryng whiche sickenes as Auctors write he caused his crowne to be set on the pillowe at his beddes heade'.

The two men became friends over the years, with Murray making frequent visits to the surgeon at Broadmoor. When Minor's mental health deteriorated, Murray campaigned for his release, and in 1910, thirty-eight years after entering Broadmoor, Minor was allowed to return to America. He died in a hospital for the elderly insane in Connecticut ten years later.

Minor outlived his friend Professor Murray, who was still only at the letter *T* when he died in 1915, working on the word *take*. The ten-volume dictionary was finally completed in 1928, almost thirty years after Murray began it. The final letter to be completed was *W*, as the team working on *XYZ* had already finished. In the *W* team, by the way, working on *waggle* to *warlock*, was one J. R. R. Tolkien.

Today, with a fully computerized, online *OED*, editor John Simpson and his team of lexicographers are coming face to face with the English language's boundlessness. 'It's constantly moving,' says John. 'It's fascinating.' New words are coined one minute and spread like wildfire the next. As a result, the rate of change in language itself has switched into hyperdrive. A printed version – with a due date estimated some time in 2037 – will be so vast, so heavy and so expensive that it's little wonder that the Oxford University Press may consider never printing the *OED* again.

The online dictionary's advanced author quotations search is tempting for any prolific writer interested in discovering how often they have been cited as a source. Type in 'S Fry' and, as well as telling you it's short for *stir fry*, the search engine throws up twenty-three words listed against the name, among them *paramnesia*, *to pinken* (taken from P. G. Wodehouse), *ack-emma* and *pip-emma* for a.m. and p.m. (Woodhouse again) and *otherwhere* (not a neologism – its first reference is in 1400!).

For a lover of words, John must have quite simply the best job in the world. Working with a living, moving stream of words, the whole of the English language his bailiwick. John is somewhat more restrained. 'It's rather nice,' he murmurs.

Libraries

As Stephen Fry exults, 'There's an excitement about the cries and whispers and the solaces and the seductions that are contained within the bound form of a single book. And when you have thousands together it's as if all of human history, all of human hope is captured and is murmuring to you like sirens, pulling you in.'

Libraries are more than just buildings, just as books are more than just print and ink. As the poet and political theorist John Milton said, 'Books are not absolutely dead things, they do contain a potency of life in them. He who destroys a book, kills reason itself.' Perhaps that's why tyrants everywhere have always attacked libraries before almost anything else. In the third century BC, the Chinese Emperor Shi Huangdi ordered that all literature, philosophy and poetry written before his dynasty should be destroyed. Books from the Persian library of Ctesiphon were thrown into the Euphrates in AD 651 on the order of Caliph Umar. And Mongol invaders in 1258 attacked Baghdad's House of Wisdom, the single largest library in the world at the time, destroying some of the oldest books ever written. It was said that for six months the waters of the Tigris ran black with ink from the enormous quantities of books flung into the river.

And it's not just in ancient times that such wanton acts of

destruction were carried out. The twentieth century saw more book burning than any other time in human history – in Nazi Germany, in Bosnia, in Afghanistan

The story of libraries is as old as the story of writing itself, for wherever and whenever we have committed knowledge to clay tablet or papyrus, bamboo or palm leaf, parchment or paper, ways of keeping that information safe and accessible have had to be found. The first libraries were mostly archive collections of commercial transactions or inventories – like the Sumerian temple rooms full of clay tablets in cuneiform or the papyrus records of the governments of ancient Egyptian cities. An archaeological dig in the ruins of a palace in Ebla in Syria uncovered shelves of clay tablets dating back to 2500 BC. The tablets – as many as 1,800 in complete form and thousands more in fragments – were inscribed with lists of the city's imports and exports, royal edicts, religious rituals and hymns and proverbs. They had been stored upright on shelves, positioned so that the first words of each tablet could be seen easily, and separated from each other by pieces of baked clay. Civilization's first library stack, perhaps.

The flourishing of literacy and intellectual life in ancient Greece saw the extension of libraries into public life. An author would write his book on nature or astronomy on a parchment or papyrus scroll, take it to be copied (by hand of course) at a copy shop and then arrange for those copies to be sold via book dealers. Some copies were of a higher standard than others so a decree in Athens called for a repository of 'trustworthy' copies. By the time of the Romans, collections of scrolls were available in the dry sections of the public baths for bathers to peruse.

The most prevalent collection of writings, however, were private ones. In the fourth century BC, the Greek philosopher Aristotle amassed a huge personal library. According to the historian and geographer Strabo, writing 300 years later, he 'was the first to have put together a collection of books and to have taught the kings in Egypt how to arrange a library'.

That library was the wondrous Library of Alexandria, built around 300 BC by Egypt's Ptolemy I. It was apparently the brain-child of one of Aristotle's disciples, Demetrius, who inspired

At its peak the Library of Alexandria stored three-quarters of a million scrolls

Ptolemy with the vision of a public library open to scholars and holding copies of every book in the world. There are a number of colourful stories about how Ptolemys I, II and III went about amassing the ancient world's greatest collection of books. Ptolemy I sent letters to kings and governors, begging them to send back to his library all the works they could get their hands on. Ptolemy III is reputed to have ordered every visitor to the city to hand over all the books and scrolls in their possession. These were then quickly copied by library scribes and handed back to the traveller; the originals were kept for the library. Legend has it that Mark Antony gave Cleopatra over 200,000 scrolls for the library as a wedding gift, 'borrowed' from the rival ancient Greek Library of Pergamon.

At its peak the library held perhaps three-quarters of a million scrolls, many of them different versions of the same text. The papyrus and leather scrolls, with wooden identity tags attached, were kept in a honeycomb of pigeonholes inside ten great Halls dedicated to different areas of learning – mathematics, astronomy, physics, natural sciences and so on.

During the second century BC, people in the Mediterranean region began using parchment, a much more expensive alternative to papyrus. One theory is that the Library of Alexandria was using so much papyrus that Egypt had very little left to export. Another suggestion is that rivalry between the two Libraries of Alexandria and Pergamon was so intense that Ptolomy V tried to cut off the supply of books to the Greek rival by banning the export of papyrus.

Whatever the reason, the shortage of papyrus meant that the library at Pergamon was forced to rely on parchment instead. Indeed parchment – *pergamenum* in Latin – derives its name from the city. It took several centuries for it to develop into a viable alternative to papyrus and become the standard writing material for medieval manuscripts. Alexandria's library lasted for around five centuries, until fires and civil war reduced and destroyed the vast collection.

The spread of Christianity and monasticism was a vital ingredient in the growth of libraries. The monasteries became the great centres of copying and preserving of codex manuscripts which had now replaced the scroll. The first Benedictine monastery, set up in Montecassino around AD 529, had a rule for monks which mandated: 'During Lent . . . let them receive a book apiece from the library and read it straight through.' Copying by hand was a labour-intensive, expensive business, and the manuscripts were prized possessions. Books were often chained to lecterns or shelves to prevent theft. Despite this, monasteries began lending volumes to each other – the first inter-library loaning system. It was these manuscripts – copied, sold or filched – which formed the core of the libraries of the Middle Ages.

The Renaissance saw a new golden age of libraries as the aristocrats of Europe vied with each other as patrons of learning and the arts by amassing large private collections of early manuscripts. The library of the Medici family of Florence was so enormous that they commissioned Michelangelo in 1523 to build the Biblioteca Medicea Laurenziana (The Laurentian Library) to house it. This period also saw the founding and growth of universities – and with it the university library. Another patron of the arts, Humphrey Duke of Gloucester, youngest brother of Henry V, donated his collection of 280 manuscripts to Oxford University in the early 1400s. 'Duke Humphrey's Library' became the foundation of Oxford's Bodleian Library.

The Bodleian Library at Oxford University is a network of thirty libraries, linked by underground tunnels

The Bodleian is one of the oldest libraries in the world, home to over 11 million books, maps and documents, a glorious institution, described by Yeats as the most beautiful building in the world (although he died before the building of the New Bodleian, which travel writer Jan Morris described as looking like a municipal swimming bath). For over 400 years scholars have had to swear a solemn oath on joining the library: 'I hereby undertake not to remove from the Library, or to mark, deface, or injure in any way, any volume, document, or other object belonging to it or in its custody; nor to bring into the Library or kindle therein any fire or flame, and not to smoke in the Library; and I promise to obey all rules of the Library.' Until the nineteenth century the oath would have been in Latin, but today one can choose from over 200 languages, from Maori and Manx to Celtic and Yiddish.

The line in the oath promising not to 'kindle therein any fire or flame' is a heartfelt one. Over the whole history of librarianship hangs the spectre of the burning of the Library of Alexandria, the greatest library of the ancient world. The ban on fire meant that the Bodleian was entirely unheated until 1845 and lacked artificial light until 1929. This kept opening hours rather short – as little as five hours a day during the darkness of winter.

The Bodleian – affectionately known as the Bod – comprises a network of thirty libraries linked by underground tunnels. It is stuffed with literary treasures – ranging from four original manuscripts of the Magna Carta and a copy of the Gutenburg Bible to the handwritten manuscript of *The Wind in the Willows* and the entire archive of Alan Bennett. And of course there's a copy of every book published in the UK, as well as printed material like the Yellow Pages, instruction manuals and every newspaper. The library is simply bursting at the seams, with an eye-popping 6 million books stored on 117 miles of shelving. It's a far cry from Oxford's first library, founded in 1320, a small first-floor room with a collection of chained books that could fit inside a single chest. The collection grew steadily, and in 1488 the Duke Humphrey's Library was built to house the gift of manuscripts from the Duke of Gloucester. The new library was plundered during the Reformation, when, according to the antiquarian Anthony Wood, 'some of those books so taken out by the Reformers were burnt, some sold away for Robin Hood's pennyworths, either to Booksellers, or to Glovers to press their gloves, or Taylors to make measures, or to Bookbinders to cover books bound by them'. By the end of the sixteenth century the university had barely a library at all. All the furniture had been sold and only three of Duke Humphrey's original books remained in the collection.

Enter Sir Thomas Bodley, a diplomat in the service of Queen Elizabeth I, who decided on his retirement to 'set up my Staffe at the Librarie dore in Oxon'. He poured money into refurbishing the library, which reopened in 1602 as 'Bodley's Library'. It's jokingly said the library was founded on pilchards, as Sir Thomas' wife was the widow of a wealthy fish merchant. Sir Thomas set about restocking the new library, donating some of the books himself and encouraging friends, amongst them Sir Walter Raleigh, to do likewise. Bodley didn't want any books in English at first – he called them 'baggage books' and 'riff raff' – but soon realized that books were increasingly being printed in English rather than in Latin. And so in 1610 he made an historic agreement with the Stationers' Company in London to put a copy of every book registered with them in the library. The arrangement set the Bodley on a road of continuous expansion and formed the basis for copyright libraries across the world. Today the Bodleian is one of six copyright libraries – or legal deposit libraries as they're now called – in the British Isles , entitled by law to receive a free copy of every book published in the UK. (The other five are the British Library, Cambridge University Library, the National Libraries of Scotland and of Wales and the Library of Trinity College Dublin.)

One of the early fruits of the deposit arrangement was the acquisition of Shakespeare's First Folio. (*Folio* is a printer's term for a sheet of paper folded in half to make two leaves or four pages of a book or manuscript.) It was sent by the Stationers' Company to Oxford in 1623, bound by William Wildgoose, a local stationer, and chained to a shelf in the Bodleian. At some point in the 1660s it was sold and replaced by the Third Folio, the librarian presumably assuming the First Folio to be out of date. He didn't anticipate it becoming one of the most valuable printed books in the world. A copy sold for £3.73 million at Christie's in New York in 2001. The First Folio was lost to the library for 240 years. When it finally resurfaced as part of a family heirloom in 1905, the Bodleian had to launch a public appeal to raise the hefty asking price of £3,000.

It was the largest sum it had ever paid for a book – the previous record payment fifteen years before was £221 and 10 shillings for a volume of Anglo-Saxon charters. Part of the First Folio's appeal is surely the wear and tear of the pages, proof that in the mid 1600s, Shakespeare's *Romeo and Juliet* was clearly the favourite of students.

As the Bodleian became home to a growing number of valuable tomes and manuscripts, the oath 'not to remove from the Library . . . any volume' was vigorously enforced. No one has ever been allowed to take a book from the Bod; it has to be read on the premises or not at all. Even Charles I, who had made Oxford the royalist capital during the Civil War, was refused when he tried to borrow a volume. Cromwell too.

Like Alice, the Bodleian's book collection grew and grew and grew – as did the number of buildings needed to house them. Today books are arriving at the rate of 1,000 a day. It's a storage nightmare.

The New Bodleian building was built in 1939. Ian Fleming kept his personal library here during the war and donated the manuscript of *Chitty Chitty Bang Bang* as a thank you. This building alone has 95 kilometres of books. There's another two and a half million books sitting in humidity-free darkness in a salt mine in Cheshire.

The problem of where to keep the majority of the least-read books is being met with the opening at the end of 2010 of a huge storage facility in Swindon, 25 miles outside Oxford. The removal of 6 million books and more than 1.2 million maps is well underway. All the books in high demand will stay in Oxford as well as all the Bodleian's literary treasures.

Among these treasures is the *Abingdon Missal*, created for the abbey of Abingdon in 1461. It's handwritten in tiny, Gothic script and illuminated on vellum. You can even see the lines the scribes ruled to keep their writing straight. This missal was one of the last hurrahs of labour-intensive, handwritten manuscripts. It was produced at almost exactly the same time that Gutenberg was printing his Bible and sounding the death knell of the copiers, although it took some years for print to be completely established.

Then there is *The First Book of the Natural History* by Pliny the Elder, translated into the Florentine language for the King of Naples, a heavily bound manuscript from 1476, the same year as the printing press arrived in England. It is utterly beautiful, the most spectacular surviving copy of the work, with a binding made from green goatskin studded with silver ornaments. It's an example of a crossover between new and old technologies: it's printed in a very modern font (Times Roman, the inspiration for the Times New Roman typeface in the 1930s) and is in vernacular Italian rather than Latin, but it uses vellum and hand illumination to make it look like the traditional manuscripts.

Some would say that world is coming to an end, and a new digital age is replacing it. Books and academic papers and maps and newspapers are now being produced electronically, and the Bodleian has a duty to preserve them all. Today racks of computer servers sit alongside the bookshelves, saving and archiving millions of gigabytes of information. It's a daunting task, but Richard Ovenden, the library's associate director and keeper of special collections, insists that good librarianship is all about ensuring that information from the present can be accessed by those in the future.

'We have staff whose job it is to keep stuff safe so that scholars in 400 years' time will be able to access the information that's being produced now, just as we're able to access the information printed by a great scholar . . . The University of Oxford has been keeping records since the late twelfth century, so it's almost genetically in our system that we need to keep this information over a long term.'

The Bodleian looks after the private papers of seven former prime ministers, which historians can pore over. The archiving of historical material will continue, says Richard, but in different ways.

'The politicians of today and tomorrow are communicating with their friends not by writing letters like Disraeli did, thousands and thousands of them, but by sending emails. They're communicating with their constituents not by publishing little pamphlets but by blogging and tweeting and putting web pages up. And as we're one of the great repositories of twentieth- and twenty-first-century politics, we're having to develop our own digital infrastructure to capture that information, keep it safe, manage it and make it accessible to researchers.'

The potential for digitalization in the library system is huge. The Bodleian has signed a deal with Google to digitize more than a million books and has already scanned about half a million of them. The benefits have already been seen, as it's opened up huge ranges of nineteenth-century books that sat unseen on shelves for 150 years. Richard Ovenden is excited about the future.

'It's always been part of what we see as "serving the whole republic of the learned", as Sir Thomas Bodley put it – sharing our knowledge. There's no point in having this great body of information if you can't give access to it.'

Andrew Carnegie

Although libraries in various forms have been around since ancient times, it was only in the nineteenth century that the truly public library, paid for by taxes and run by the state and freely accessible to everyone, came into being on a large scale. Before that, libraries were predominantly the private collections of individuals, or belonged to religious or educational institutions, and weren't open to the general public.

The Victorian era was the era of self-improvement and education of the masses, and public libraries played a central role. As did a man whose name became synonymous with the movement: the philanthropist Andrew Carnegie.

Carnegie emigrated to America from Scotland and became one of the richest men in the world. A sort of nineteenth-century industrial tycoon version of Bill Gates, he spent forty years amassing a huge fortune from iron and steel, and then started to give it away for the public good. 'The man who dies thus rich, dies disgraced,' he wrote and, good as his word, by the time he died in 1919, he had poured more than £70 million into educational and scientific ventures, including the building of nearly 3,000 free libraries, principally in the USA and Britain.

It's an archetypal Victorian rags-to-riches story and a classic tale of the American dream. Born in 1835 in Dunfermline in Fife, Carnegie was the son of a weaver. The family was poor, and Andrew's father struggled to find work when the steam-powered loom arrived and took business away from the

*Philanthropist Andrew Carnegie supported the
building of more than 3,000 public libraries*

hand-loom weavers. So they took a gamble, borrowed money
for the passage, and father, mother and two sons emigrated to
America in 1848 and settled in Pittsburgh, Pennsylvania.

From that day forward, twelve-year-old Andrew worked and

worked and worked, pulling himself up by the bootstraps through a series of jobs, starting as bobbin boy in the same cotton factory where his father found work. Then he became a messenger boy in the Pittsburgh telegraph office, doing everything he could to get ahead, including memorizing the city's street lay-out and the names and addresses of the important people he delivered to.

The only formal education Andrew Carnegie had was three or four years at a small school in Dunfermline, but throughout his life he was passionate about education. He attended night classes when he could, and when he delivered telegrams to the theatre he used to stay on and watch the plays.

And he read. Books were his passport to educating himself and making his way in the world. In his autobiography he paid tribute to a Pittsburgh gentleman called Colonel Anderson, who had a private library of about 400 books which he would open to local boys every weekend and let them borrow any book they wanted. This generosity stayed with Carnegie throughout his rise to power and wealth, and inspired his own library philanthropy. 'Only he who has longed as I did for Saturdays to come,' he wrote later, 'can understand what Colonel Anderson did for me and the boys of Allegheny. Is it any wonder that I resolved if ever surplus wealth came to me, I would use it imitating my benefactor?' He gifted his first library to his home town of Dunfermline. Carnegie gave £8,000 for the building, his mother laid the foundation stone, and it opened in 1883. It was an instant success, and by the end of its first day it had issued more than 2,000 books. This was the pattern that Carnegie followed through the years: he would build and equip the library, but the deal was that the local authority had to match that by providing the land, and coming up with the budget for operation and maintenance. He wasn't just handing out money willy-nilly; he wanted the state to take on the library after he made it possible. And as a further condition to funding, local councils had to adopt the Public Libraries Acts, which, from 1850, gave local councils the power to establish free public libraries. To Carnegie this was practical philanthropism, not charity: 'I choose free libraries as the best agencies for improving the masses of the people,' he said, 'because they only help those who help themselves. They never pauperise, a taste for reading drives out lower tastes.'

He continued to endow libraries across America and all over Britain and beyond: $500,000 in 1885 to Pittsburgh for a public library; $250,000 to Allegheny City for a music hall and library; $250,000 to Edinburgh for a free library. In total Carnegie funded some 3,000 libraries, more than 600 in Britain and Ireland, and nearly three times that number in the US.

Not bad for a boy from Fife.

Penny Dreadful to Thumb Novel

'Make 'em laugh, make 'em cry, make 'em wait'
(Wilkie Collins)

Wander down any street in Japan, travel on public transport there, and you'll notice something remarkable: The thumb tribe – *oyayu-bizoku* – are at work: Japanese teenagers typing constantly on their mobile phones, communicating in text and, most recently, reading and telling stories in the form of so-called *thumb novels*. Lurid romantic fictions are tapped out in instalments by novice authors and devoured by mostly young and female mobile-phone users.

It's a form of writing which sits happily with the publishing phenomenon of serialized fiction, which, arguably, marked the beginning of society's first mass popular culture. The popularity of the novel in instalments amongst the working classes went hand-in-hand with developments in printing and an increase in mass literacy rates in the nineteenth century. All the great Victorian novelists, including George Eliot, William Thackeray and Joseph Conrad, published their newest works of fiction in instalments, but the undoubted genius of the format was that prodigious wordsmith Charles Dickens.

Most of Dickens' working-class readers couldn't afford the price of a full-length novel, so he published the bulk of his major novels in monthly or weekly instalments which could be bought for a shilling. Many of the instalments ended with a cliff-hanger, ensuring the sales of the next one. Sales of copies of his first serialized novel, *The Pickwick Papers*, in 1836 went from 1,000 for the first edition to 40,000 for the final instalment. Brimming with memorable characters, strong narrative, acute observation, comedy, villainy and tragedy – all the Dickens trademarks – *The Pickwick Papers* became one of the most successful novels of its time and thrust the twenty-four-year-old author into literary stardom.

Charles Dickens' The Pickwick Papers *was serialized and each edition sold for just a shilling*

Dickens' writings appealed to people across the social spectrum, and due to the new technological advances in publishing and transport he could reach a reading public of unprecedented size. His serialized novels were so gripping, so entertaining, that when he published *The Old Curiosity Shop* in instalments between 1840 and 1841, weekly sales rose to 100,000. As it reached its climax – the death of Little Nell – Dickens was inundated with letters begging him not to kill her off. It's said that when English ships docked in New York, crowds filled the piers, shouting 'Is Little Nell dead?' Nell's death scene caused even grown men to cry. Dickens' associate, the Scottish judge Lord Jeffrey, was apparently found openly weeping and confessed, 'I'm a great goose to have given way so, but I couldn't help it.'

Penny dreadful stories were full of violence, gore and crime

Cheaper alternatives to the serial story began to be published in England around the same time Dickens was writing. The 'penny dreadful', or 'penny blood' or even 'penny awful' were serialized stories of sensational fiction, full of violence, gore, crime and horror, vampires,

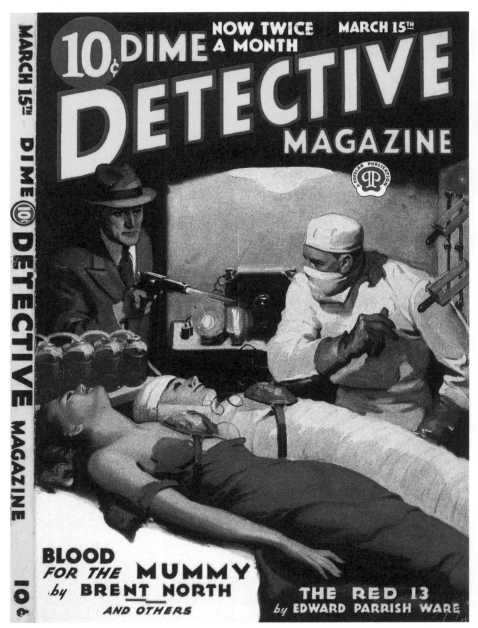

The dime novel was the American equivalent to the penny dreadful

monsters and ladies in distress. The infamous Sweeney Todd, the Demon Barber, made his literary debut in a penny dreadful serial called *The String of Pearls: A Romance* in 1846. Printed on eight pages of cheap paper and sold for a penny, the bloods were aimed at working-class adults, although eventually the readership became almost exclusively boys.

The dime novel was the penny dreadful's equivalent in the United States. The adventures of the American frontier characters such as Calamity Jane, Deadwood Dick and Wild Bill Hickok were the precursors of the first Western novels. The penny dreadfuls were formulaic and lurid, churned out by a string of mostly hack writers – what we might call pulp fiction today – but they offered a kind of written storytelling with a deep, widespread appeal. English writer G. K. Chesterton explained the attraction: 'My taste is for the sensational novel, the detective story, the story about death, robbery and secret societies; a taste which I share in common with the bulk at least of the male population of this world.'

If the serialized novels of Dickens and the penny dreadfuls are the television of the nineteenth century, then what are we to make of the latest innovation in novel writing of the twentieth-first century?

Take a subway in any Japanese city and you'll notice most of the young people glued to their mobile phones. They're not making phone calls – Japanese etiquette frowns upon talking on public transport; they may be text messaging; but an awful lot of them will be using their flip-top mobiles to compose, share or read a new literary art form – the *keitai shousetsu*, or mobile-phone novel. In a country where an entire generation has grown up using mobile phones to communicate, watch TV and films, shop and surf the internet, writing thumb novels on the keypad is the obvious next step. Just as Dickens published his novels in instalments, so the thumb novel chapters, often only 100 words long, are sent directly to the reader one by one. They're downloaded from mobile-phone novel sites, the largest of which, Maho i-Land, has 6 million members and more than a million titles. The stories are mostly romantic fiction with racy storylines involving rape, pregnancy and, of course, love. The thumb novels are often in diary or confessional form, heavy on dialogue and short paragraphs and liberally sprinkled with slang and emoticons. They tend to be written by novices and read by teenage girls and young women – although the first one of its kind was actually thumbed by 'Yoshi', a Tokyo man in his mid-thirties. 'Yoshi' (thumb novelists nearly always use pseudonyms) set up a website in 2000 and began posting instalments of his novel *Deep Love*, a lurid tale of prostitution, Aids and suicide. It was so popular that it was published in book form, selling 2.6 million copies; a spin-off TV series, manga and movie have followed.

One of the most famous thumb novelists is an eighty-nine-year-old Buddhist nun called Jakucho Setouchi. She's something of a national treasure in Japan, famed for her translation of what's thought to be the world's first novel – the 1,000-year-old *The Tale of Genji*. Setouchi was appearing as a judge at the annual Keita Novel Awards when she suddenly made an announcement. For the last few months, she said, she had been posting a thumb novel called *Tomorrow's Rainbow* under the pen-name 'Purple'. Setouchi explained that she'd tried to write her simple love story on her mobile phone but found the thumbing too difficult. In the end she used paper and a fountain pen and sent *Tomorrow's Rainbow* to her publisher for conversion into text.

'I'm an author,' she told her audience of thumb novelists. 'When you finish a novel, to sell tens of thousands would be a tough thing for us, but I see you selling millions. I must confess that I was a bit jealous in the beginning.' Many of the popular online novels are published in print, in the first half of 2007, five of the novels in Japan's bestseller list started as thumb novels. The novels are not great literary works. They lack scene setting or character development. Yet, however crude, the thumb novel genre has encouraged people in their tens of thousands to tell stories, and millions of others to read them. That's something Dickens himself would have applauded.

Death of the Book?

There's a right royal battle of words raging over the future of the book – or more specifically, the future of the printed book. For, contrary to the doomsayers who have warned of the death of reading, people are in fact reading in increasing numbers; just in different ways. The digital revolution has brought us the e-book, the electronic book, and many lovers of the printed word aren't happy. The American novelist the late John Updike was addressing a gathering of booksellers about the future of the printed book. 'The book revolution,' he told them, 'which from the Renaissance on taught men and women to cherish and cultivate their individuality, threatens to end in a sparkling pod of snippets.' Updike was bemoaning the tidal wave of unedited, often inaccurate, mass of information on the internet. 'Books traditionally have edges,' he said and concluded with a bibliophile's call to arms. 'So, booksellers, defend your lonely forts. Keep your edges dry. Your edges are our edges. For some of us, books are intrinsic to our human identity.'

Updike's impassioned plea was met with an equally emotional response from e-book lovers. 'Print is where words go to die,' wrote one.

Professor Robert Darnton, a leading scholar in the history of the book and director of the Harvard University Library, insists that the book is very much alive, and the statistics seem to prove it. Each year more books are produced than the previous year. There was a dip during the recession, but next year, he says, 1 million new titles will be produced worldwide. And yet, the era of the book as the unrivalled source and vehicle for knowledge is undoubtedly coming to an end. We're living in a time of transition in which the two media co-exist. It is, says Professor Darnton, a most exciting time.

'One thing we've learned in the history of books is that one medium does not displace another. So the radio did not displace the newspaper. And television did not kill the radio. And the internet did not destroy television, and so on. So I think actually what's happening now is that the electronic means of communication, all kinds of handheld devices on which people read books, are actually increasing the sales of ordinary printed books.'

So more people are reading; or it could be that the same number of people are reading more. Probably a bit of both. What's certain is that a lot of people use handheld electronic devices for one kind of reading – travelling, research, instant literary gratification – and a printed book for another kind of reading – in the bath, snuggled in bed, swinging in a hammock. The one mode is fast, useful, functional; the other is a more physical, emotional engagement.

Professor Darnton agrees: 'I'm very attached to books and to manuscripts because I spent many years reading archival sources. The feel of the paper. The way the ink blends into it. The excitement of opening a new carton of manuscripts in the Bibliothèque Nationale. Undoing the ribbon, folding it back, seeing what's going to be inside the folders. Coming into contact with vanished humanity. There's nothing like it. It's thrilling. And I think for people who are not professional historians, but are readers, there is a similar thrill. You crack open a book and get the spine to work in such and such a way. I know I'm sounding sentimental or romantic, but it's something that is, I think, profoundly part of us, those of us who were brought up in this way. And I would say that about newspapers as well. None of my students read printed newspapers. None. They all get their news online. But for me, the front page of the *New York Times* is a map of yesterday. That sensation of beginning the day looking at the map of yesterday, is something I think that is missing in a lot of things.'

Attachment to the printed book is not just the preserve of an older generation. According to one survey, 43 per cent of French students who were asked why they didn't use e-books said they missed the book smell. There's even a jokey aerosol 'e-book enhancer' called 'Smell of Books' on the market, promising to 'bring back that real book smell you miss so much'.

Nevertheless, there is a growing number of people for whom the book means very little. In fact, apart from in school, they may never read a book. It doesn't mean they don't read, simply that they never open a physical book. Electronic books have all sorts of advantages. In parts of the world where schools and libraries can't afford books, accessing a database of information in cyberspace is a cheaper alternative. Archives can be accessed with the press of a button – with one touch you can be inspecting the writing on a Sumerian clay tablet; another touch and you're gazing at a fifteenth-century illuminated manuscript. Unlike the printed book, e-books can be updated and corrected constantly; readers can be linked to related subjects via live footnotes. You can even scribble your notes electronically on the margins and then share them.

Professor Darnton recognizes the huge possibilities of the digital media. He recently published a history book which includes songs that can be accessed via the internet and listened to while you're reading the book. 'So we're taking in history through the ear as well as through the eye.' And what of the printed book, apart from its smell? At a practical level, we only need our eyes and hands to read it: no electricity, batteries or internet access necessary. We like to own books, to collect them, to display them on our shelves. Our favourite books have bits of ourselves in them. It may be that at the end of this transitional period when print and digital co-exist, the printed book will become the preserve of the few. Those who like to turn a physical page are already able to go to a shop, order a digitized text, and it's printed, trimmed and bound within four minutes. As someone (probably a publisher) once said, as long as authors have a mother, they'll want to give her a copy of their book.

Message to the Future

Some very big questions are being asked about the future of written language – not in the next few centuries but in the next tens of thousands of years. Given the difficulties we have in reading the Old English of *Beowulf*, let alone in decoding the early writing systems of 5,000-year-old civilizations, how can we make our language readable to peoples of the future? The American government has been wrestling with the problem of how to warn future generations about the spent nuclear fuel and high-level radioactive waste which they have been burying in underground chambers deep in the Mojave Desert. The waste will remain dangerous for at least 10,000 years. No one knows which languages, if any, will be spoken by AD 12,000, give or take a year or so. So how will we warn the future about the risks?

In 1984 the late linguist Thomas Sebeok was asked to come up with suggestions for the most effective 'keep out' signs. Sebeok argued that, in 10,000 years, all the world's spoken and written languages would likely have faded away. So he recommended the use of warning signs and symbols and the creation of what he called an 'atomic priesthood' of scientists and scholars who could pass down the knowledge about the danger of the site from generation to generation. These priests would also be tasked with passing on 'artificially created and nurtured ritual-and-legend' myths about evil spirits so that people would be too scared to go near the site.

In the 1990s, a team of linguists, scientists and anthropologists was commissioned to design the warning system for the Mohave Desert nuclear dump site. The art designer on the team is Jon Lomberg, one of the world's leading space artists, who illustrated most of astronomer Carl Sagan's books. Lomberg has created some of the most durable and far-flung objects ever produced by a human being. His design for the aluminium jacket cover of the *Voyager* spacecraft's Interstellar Record message from the people of Earth is intended to last for a thousand million years. This is definitely a man who knows a thing or two about messages for the future. And trying to talk to aliens has given us some ideas about how best we might create meaningful messages for the generations 10,000 years ahead of us.

Lomberg explains how straightaway the Mohave Desert design team rejected the idea of a magic bullet – one perfect way of solving the problem. They decided on a shotgun approach, looking for a variety of options. Their first thought was to go for picture symbols.

'People change, culture changes; what we looked for was what was invariant. In the case of languages we know that languages evolve and have a half life. For most of us Shakespeare can be a little tricky to read, Chaucer difficult and *Beowulf* impossible. And that's just one thousand years. Pictures, on the other hand, do seem to be universal. We can recognize the animals on the walls of the caves, and one of the common motifs that you find in art all over the world is the stylized picture of a human. A stick figure . . . You can tell it's a person and you can tell if it's standing up or lying down and what they're doing – running, throwing a spear, sitting on a throne . . . Symbols are a sort of a mid-point. A symbol in a sense is a word that doesn't need translation.'

But Lomberg and his team discovered that there are actually very few universal symbols. Carl Sagan wrote to the team to suggest that the skull and crossbones be used as the symbol of danger and

Mr Yuk. Will language as we know it no longer exist in the future?

death. Yet it changes in meaning from culture to culture. In the Middle Ages it represented the resurrection of Christ and symbolized hope. It was only when it was put on gravestones that it began to be associated with death and later with the Jolly Roger pirate's flag.

The team decided to use double symbols – the skull and crossbones plus the Mr Yuk figure, a frowny face with a tongue sticking out which is widely used in the United States as the symbol for poison.

'The other thing that seems to be universal is the notion of a storyboard,' says Lomberg, 'a series of drawings that show a series

Comic Strip 1   Comic Strip 2

of events in time. We find it in the Bayeux Tapestry, in the scrolls of Japan that depict the invasion by the Mongols.' Lomberg started his art career drawing comics and was keen to explore the idea that the symbols could be defined within a comic strip story. He believes that, although comic strips are without language, like a Charlie Chaplin movie their language is universal. Crude cartoons were drawn, each picture cell stacked from the top down since, although not all cultures read from right to left, they all read from top to bottom.

Comic strip 1 shows a person removing something from a container with the poison symbol on it; the poison symbol appears on their chest and they fall on the ground and die. Comic strip 2 is a warning about the long-term effect of the poison. A child is shown going into a container that's marked with the radiation symbol. The same person is next seen as an adult – we know time has passed because the tree in the background has grown tall – and then he's falling down. The symbols, comic strips and warnings in the six official languages of the UN will be etched into a series of markers – granite pillars erected around the nuclear waste site. Lomberg and his team have until 2028 to submit their final plan.

Lomberg and a number of the team members have a background in SETI, the Search for Extra Terrestrial Intelligence. They helped choose the messages carried by the *Voyager* spacecraft into outer space in 1977, including recordings of greetings in fifty-five languages. Impossible to know what an alien will make of 'Hello from the children of Planet Earth', but, as Carl Sagan pointed out at the time, perhaps the most important impact of the message will be that we cared enough to try to communicate.

CHAPTER 5

The Power and the Glory

Language is our supreme evolutionary achievement, and this chapter is about how we can use it to sublime effect. We save our best words for our most complex thoughts, our attempts to win a sexual mate, deal with the inevitability of our own death or to persuade someone to do something they don't want to do. It's where language blooms in resplendent fashion. That doesn't mean it has to be florid; it can have the simplicity of a haiku. Three hundred years ago, Alexander Pope wrote: 'True Wit is Nature to advantage dress'd, What oft was thought but n'er so well expressed.'

A manual on how to hunt, plough, build a house, fight a war or paddle a boat can get by with pretty basic language. But courtship won't go far if you say simply, 'You look nice, let's have sex.' The same goes for death: 'What, old dad dead?' is a great line from *The Revenger's Tragedy*, but mainly for dramatic impact. Hamlet's 'Oh that this too too solid flesh would melt, thaw and resolve into a dew!' certainly has more eloquence to it. Is it better? Well, that's very subjective, and we all have our favourites, be they poems, songs, plays, novels or the speeches of a gifted orator.

The acme of language is literature; it defines our species, gives us voice, personality and history. Quite simply, it tells our story.

Back in Turkanaland

Essentially, what we like as a species across all cultures and throughout our history is a good story, well told: as Joyce would have it, the right words in the right order. We return first to that village in Turkanaland in north-east Kenya we visited earlier in the book, for the Turkana can tell us something about the beginnings of storytelling.

The village is about ten miles from the Sudan border. The Toposa, mortal enemies of the Turkana, live on the other side. The Toposa are as obsessed with cattle raiding as the Turkana, and for the young men of both tribes the raids are a way of life: ritually important and essential as a means of getting cattle to pay the bride price, so they can get married. It's also tremendously exciting. Sometimes as many as a hundred heavily armed Toposa will cross the border and launch attacks. Up until thirty years ago both Toposa and Turkana relied mainly on ancient muskets, Enfields and spears, but now Kalashnikovs have replaced them. There is not a single spear to be seen,

Turkana warriors keeping their traditions alive

and even young boys of fourteen have an automatic rifle slung casually over their shoulders. As the day wears on the warriors start to drink their millet beer. Drunkenness and guns are never a good combination.

But what the alcohol does is lower the inhibitions of the young warriors, and they start to tell stories of their escapades against the Toposa. Soon a crowd gathers and, emboldened by their audience, the young men launch into a full re-enactment of their latest raid. Miming their ambush of the Toposa warriors, they tell the story of capturing a hundred head of cattle, losing a couple of their men in a firefight and then the Toposa taking back some of their cattle only to be re-ambushed. The crowd love it, as do the warriors. This act of mimesis, miming out their story – with a few embellishments – goes to the heart of storytelling. There's also a plot, action, humour (one of the men got shot in the buttocks – very Forrest Gump) and resolution.

The story is an improvised bit of drama without any religious or moral undertones. It's a bit like an action buddy movie – a platoon of warriors on a dangerous mission meets with adversity, overcomes it; some are wounded, but the end is a happy one because the warriors return home victorious.

The Turkana have other stories, many of them traditional myths which attempt to answer the major themes of their lives. They do have a creator figure, Akuj, and there are many stories in which he is the protagonist. Most involve rain, which is the life-or-death commodity for the Turkana. These stories, like the Judaeo-Christian Bible, try to reconcile the big issues that trouble the Turkana. In the harsh environment where they live, the Turkana believe that the unreliability of the rains which cause drought, famine and death can be alleviated by the intercession of Akuj, the bringer of rain.

Storytelling is as old as language and is rooted in our life as a social animal. It is a universal across all cultures throughout history. When the Turkana tell a story they do more than just entertain the community, they bind it together in a communal ritual. This ritual (and it's not so different from being read a Harry Potter book, or going to the cinema to see *Avatar* or watching a performance of *Hamlet*) illustrates two of our most important traits: our ability to create imaginary worlds – fantasy – and our capacity to empathize. Empathy allows us to reach out to others and understand another's emotions and needs whilst fantasy allows us to postulate alternatives and imagine new ways of seeing reality. Both are at the heart of good storytelling. But what could be the evolutionary advantage of being so prone to fantasy?

'One might have expected natural selection to have weeded out any inclination to engage in imaginary worlds rather than the real one,' says linguist Steven Pinker, but he thinks stories are in fact an important tool for learning and for developing relationships with others in one's social group. And most scientists agree: stories have such a powerful and universal appeal that the neurological roots of both telling tales and enjoying them are probably tied to crucial parts of our social cognition. Pinker's hypothesis is that as our ancestors evolved to live in groups, so they had to make sense of increasingly complex social relationships. Living communally requires knowing who your group members are and what they are doing and if possible what they are feeling and thinking. Storytelling is the perfect way to spread such information.

Songlines

Storytelling is at the heart of the indigenous people of Australia. There were over 400 distinct Aboriginal clans when the first British settlers arrived in Australia, and each of them had their own body of stories that describe and link the eternal mythical world of what they call the 'Dreaming' with the everyday world of living. The stories are called 'Songlines'.

It's almost impossible for a non-Aborigine to understand the idea of the Dreaming which underlies the Songlines. The Dreaming is an English word that attempts to describe the entire mystical and onto-logical life of the Aborigines and encompasses how they are linked to the contours of the landscape in which they live. Aboriginal actor Ernie Dingo tries to explain: 'When you talk about it, you think about it in the back of your head – "Now hang on a sec, how did that story go?" – and that moment of thought would come like a dream. So it's going back into the past through memory, rather than in the English sense, through the history books.'

The Dreaming is an extraordinarily complex body of stories that guide the Aborigines in their quest for water, hunting, fertility. Each of the Aboriginal tribes has its own stories, and these are sung in Songlines or Dreaming Tracks, the paths created by the ancestors during the Dreaming as they criss-crossed the country. This is one from the Grampian Mountains of Victoria.

The Gariwerd Creation Story

In the time before time began Bunjil, the Great Ancestor Spirit, began to create the world around us: rivers, mountains, forest, desert. He created the animals and plants. He appointed the Bram-bram-bult brothers, the sons of Druk the frog, to finish the task of naming the animals and making the languages and laws. At the end of his time on earth Bunjil rose into the sky and became a star, where he remains.

There was a giant ferocious emu named Tchingal, who ate people. His home was in the scrub, where he was hatching a giant egg. One day, Waa the crow flew past and, feeling peckish, started to eat the egg. When Tchingal came back and saw this he was furious and chased Waa all over the place and each time he escaped Tchingal crashed into mountains creating the gaps in the rocks and the streams than run through them.

So do Aborigines sing their stories because songs tend to be more memorable?

Ernie explains, 'It's both a rhyme and rhythm, and the rhythm is the heartbeat. You sing this, your heartbeat can tell you how much time you've got to travel. So the song that you sing, if it's a good old slow song, you know that this is a long journey, before you get to the next part of the verse. The story will stay similar to the rhythm of the walk and it tries to get your heart to pace you so that your step paces you to the next location in the song.'

AN AUSTRALIAN FEAST.

Songlines explore themes of hunting and fertility and each Aborigine tribe has their own set of stories

The Songlines are a brilliant, practical way of navigating. They tell you compass points, give you landmarks (depressions in the land, for instance, are remembered in the songs as the footprints of the creator beings), and reinforce the identity of each tribe by making everyone learn by heart their collective history. They also, as any parent will know who's tried to coax a child on a long walk, provide sufficient distraction and entertainment to make the journey more bearable.

Homer

If you like a gripping story, packed with adventure and heroic exploits, jealousy and loyalty, friendship and family, love and loss, and the yearning to find the way home, then read – no, listen to – this:

Tell me, Muse, of the man of many ways, who was driven
far journeys, after he had sacked Troy's sacred citadel.
Many were they whose cities he saw, whose minds he learned of,
many the pains he suffered in his spirit on the wide sea,
struggling for his own life and the homecoming of his companions.
Even so, he could not save his companions, hard though
he strove to; they were destroyed by their own wild recklessness,
fools, who devoured the oxen of Helios, the Sun God,
and he took away the day of their homecoming. From some point
here, goddess, daughter of Zeus, speak and begin our story.

It's the opening lines to the *Odyssey* (from a translation by Richard Lattimore), the towering epic poem of the trials and tribulations of the Greek warrior Odysseus, as he tries to sail home to Ithaca after the fall of Troy.

Homer's *Odyssey* may be Western literature's first – and some would say most influential – work but it was not a *written* creation. It was born out of, and shaped by, an ancient oral tradition which memorized and passed on stories, cultural values and information from generation to generation, long before we learned to write. It's more song than words.

Except that, at some point, it *was* written down. We're not sure when and we're not sure by whom: probably some time around the late eighth century BC, by a man we call Homer, but who might have been more than one person, according to scholars. Around the same time, a few decades earlier, probably, Homer's other masterpiece, the *Illiad*, was also written down, and together they form the bedrock of the Western literary tradition.

Homer's epics are great yarns, wonderful stories and, far more, they create vivid worlds of complex human desires and contradictions, where people love and suffer, fight and die, live with dignity or dishonour, struggle against misfortune and tragedy and fate. The characters try to understand the world they live in, the physical and the unseen, against a backdrop of heroic deeds and all-too-human gods. Like all the best tales, they're in essence stories of the human condition, and their depictions of the ancient world, their plots, styles, literary devices and imaginative sweep have soaked into our culture, language and art. Every epic journey, road movie, every tale of the returning warrior, every story of the absent father and the coming-of-age son, Dante's *Inferno*, the epic poetry of Milton and Pope, the classical poets, the Romantics, James Joyce's *Ulysses* – they all owe their inspiration and origin to Homer's *Odyssey*.

The Greeks themselves considered the *Illiad* Homer's greatest work; and it was the story of Achilles, and not the wanderings of Odysseus, which Alexander the Great took with him as bedtime reading on his campaigns. Both poems draw their inspiration and material from the Trojan War myths. You know the story: Helen seduced by Paris and whisked off to Troy, the face that launched a thousand ships, Agamemnon leading the Greeks against King Priam in the ten-year-long war, the burning towers of Ilium, Achilles dragging the body of Hector round the walls of the city, the Trojan horse and the fall of Troy. And then one of the Greek generals, Odysseus, gets lost on his way home. It's the ancient world's equivalent of the greatest road movie ever – except it was by boat.

It's the *Odyssey* which has given us some of the greatest stories and adventures in literature, as Odysseus and his men meet one obstacle and disaster after another: navigating between the lethal whirlpool and rocks of Scylla and Charybdis; being captured by the one-eyed Cyclops Polyphemus, whom Odysseus blinds with a stake so that he and his men can escape clinging to the undersides of the Cyclops' huge sheep.

Things are always going wrong, working against Odysseus getting safely home, either because of the gods or the stupidity of the men, or the weather. Like when Aeolus, the master of the winds, gives Odysseus a leather bag containing all the winds, except the west wind, to speed him home safely. The sailors think the bag's got gold in it, and they wait until Odysseus is sleeping and then open it. All

Homer's Odyssey *may be Western literature's most influential work*

Homer wrote epics full of vivid worlds and complex human desires

the winds fly out in a fury, and the resulting storm drives the ships back the way they had come – just as Ithaca had come into sight. Then there is the episode when Odysseus decides he wants to hear the irrisistible song of the Sirens, who would lure sailors to their death on the rocks. He gets his crew to plug their ears with beeswax and tie him to the mast, with strict orders not to untie him, no matter how much he begs. So there he is, writhing against the creaking mast, straining towards the singing

So they sang, in sweet utterance, and the heart within me desired to listen.

338

And so it goes on, twenty-four books and 12,000-odd lines of pacy, vivid poetry which have burrowed deep into our Western collective imagination in the same way as the Bible stories have done.

The mystery of who Homer was and the scholarly debates about the exact origins of these epic poems only add to the allure. There's no trustworthy information about the life and identity of Homer which have come down to us. All we have is what we read that the ancient Greeks believed, people like Herodotus, who thought Homer lived about the ninth century BC, and Aristarchus of Alexandria, who offered a much earlier date – he believed Homer lived about 140 years after the Trojan War (which we date around 1200 BC). The Greeks believed Homer was blind, and some thought he came from Chios, others from Smyrna. They also assumed that he was a poet who wrote. There was disagreement about when the poems assumed their final shape, and whether different poets wrote the *Illiad* and the *Odyssey* – but everyone, all the way down to Alexander Pope in the eighteenth century, assumed that Homer was a poet who composed with the written word.

Scholarship over the next couple of centuries challenged this traditional view and located the poems in a pre-literate culture, with an oral composition and transmission span of generations, until they were finally written down, probably in the eighth century BC. Academics argued that Homer was, in fact, one of a long line of *oral* poets and that the style of oral compositions is very different from written ones. They analysed the texts and identified the repeated use of descriptive phrases – what we call Homeric epithets – like 'divine Odysseus', the 'wine-dark sea', 'grey-eyed Athena', 'rosy-fingered Dawn', and even whole repeated lines and standard set-pieces of description. They argued that these formed a vast repository of word-groups, a sort of poetic diction, which the oral poet would draw on. He would hear them and learn others from other poets, and during the live performance of his poem, which could go on for hours, he could use them when he improvised. He could draw on words and phrases which would fit the metre and rhyme, signpost the characters for the listeners, give shape and pace to the whole. The argument continued that this kind of poetic diction could only be the cumulative creation of many generations of oral poets over centuries, and was so complex that it couldn't have been the work of a single poet.

No one today would claim that one man called Homer sat down and wrote the *Illiad* and the *Odyssey* at one sitting. But we can say that one man probably perfected what generations worked at, a magnificent poet who, perhaps over a lifetime, gathered the treasures and resources of an ancient traditional art, shaped and polished old stories, and created something vivid and new – these powerful dramas about the tragedy of Achilles and the desperate wanderings of Odysseus, which have stood the test of time and fired the imagination of succeeding generations. Whether this poet was called Homer or not doesn't really matter. Because, whatever the origins and authorship of these epics, it's their supreme storytelling verve, their poetry and the imaginative power driving them which still speaks to us.

Here's another example, from the opening of the *Illiad*, in a modern, vernacular translation by the American scholar Stanley Lombardo:

> *Sing, Goddess, Achilles' rage,*
> *Black and murderous, that cost the Greeks*
> *Incalculable pain, pitched countless souls*
> *Of heroes into Hades' dark,*
> *And left their bodies to rot as feasts*
> *For dogs and birds, as Zeus' will was done.*

It's rousing stuff, with the immediate promise of action and terrible deeds, and that ringing phrase which instantly turns the spotlight on the brooding, proud figure at the centre of the drama – Achilles and his 'black and murderous' rage.

Stanley Lombardo is so passionate about the poetry and the music of the *Illiad* and the *Odyssey* that he travels around performing them to live audiences, just as the bards did millennia ago. Armed only with a drum, which he beats at strategic points in the action, he recites sections from the poems. 'You experience the poetry of Homer in a different way,' he says. He thinks that Homer came to this level of composition through performance.

The Illiad *is full of passionate, rousing language*

The Seven Basic Plots

What makes a good story? Compulsive characters, scintillating dialogue, wonderful writing – certainly – but for most of us, the plotline is the vital ingredient in a story, the roadmap, as it were, for characters.

Most writers will tell you that there are only so many storylines to be told. Some have even tried to put a number on it. Rudyard Kipling is thought to have had a list with sixty-nine basic plots; others have argued for thirty-six, twenty and – in the case of Professor William Foster-Harris – three. He suggested happy ending, unhappy ending and the 'literary' plot, 'in which the whole plot is done backwards and the story winds up in futility and unhappiness'. Ronald Tobias argues that all plots can be boiled down to two – 'plots of the body' and 'plots of the mind'. Some have even managed to squeeze all the stories in the world into one plotline: exposition – rising action – climax – falling action – denouement.

The most useful list – certainly the best fun at dinner parties – is probably that of journalist Christopher Booker, who proposes that all narratives in the world are variations of a basic seven plots. Here they are, along with some story suggestions to get the conversations going:

Overcoming the Monster: the hero confronts and defeats a life-threatening monster or evil force. The hero returns home victorious. *Beowulf*, 'Jack and the Beanstalk' and *Dracula* would fit into this category, along with all the James Bond films, *High Noon*, *Jaws* and *Alien*.

Rags to Riches: a commonplace character, often in wretched circumstances, achieves wealth, status, beauty, happiness. 'Cinderella', 'The Ugly Duckling', 'Aladdin' (and a host of other fairy tales) have common links with *Jane Eyre*, *David Copperfield* and *Charlie and the Chocolate Factory*.

The Quest: the hero sets out on a hazardous journey to reach his goal, confronting dangers and temptations along the way. The *Odyssey*, *Pilgrim's Progress*, *Don Quixote*, *Lord of the Rings*, *Raiders of the Lost Ark* and *Watership Down* (yes, rabbits *can* be heroic).

Voyage and Return: the hero leaves home to explore another, often magical, world and, after a dramatic escape, returns to the familiar world. *The Chronicles of Narnia, Alice in Wonderland*, 'Goldilocks and the Three Bears', 'The Rime of the Ancient Mariner', 'Sleeping Beauty', *The Wizard of Oz* and *Gulliver's Travels*. Many of the Quest stories fit this bill as well.

Bonnie and Clyde, *a modern tragedy*

Rebirth: the hero is overcome by dark forces and then redeemed, often by the power of love. 'Snow White', *Silas Marner, It's a Wonderful Life, A Christmas Carol, Star Wars, The Grinch.*

Comedy: a chaos of misunderstanding which eventually

resolves itself into a happy ending. Everything from Oscar Wilde plays and Jane Austen novels to Feydeau farces and most of Shakespeare's comedies.

Tragedy: a flawed character spirals down into evil and inevitable death or disaster. From *Macbeth* and *King Lear* to *Bonnie and Clyde* and *Madame Bovary*.

There are other genres, of course – mystery, romance, sci-fi – but most of the stories will fit into the framework of one of these seven basic plots.

William Goldman

Why is it that some books have us staying up all night to finish them and others stay unread after the first few pages? What is it about one story which has us quivering for more and another which has us wriggling with boredom? The secret of what makes a good story is the Holy Grail of writers, publishers and movie and TV executives. Get it right, and you've got a chart-topping book or film on your hands. Get it wrong, and you've got a flop. Nowhere is the secret of a good story more hungrily sought than in the movie business, where hundreds of millions of pounds are made or lost at the box office.

William Goldman, double Oscar-winner and considered by many to be Hollywood's pre-eminent screenwriter, should know a thing or two about the ingredients for a good story – or perhaps not, as his most famous quote is 'Nobody knows anything.' Goldman described in his memoir, *Adventures in the Screen Trade*, how one of the highest-grossing films of all time, *Raiders of the Lost Ark*, was turned down by every studio in Hollywood except Paramount. And *Star Wars* was passed over by Universal. 'Nobody – not now, not ever – knows the least goddamn thing about what is or isn't going to work at the box office.'

It's hard to believe that, after more than fifty years in the story business, Goldman doesn't have an inkling of what works and doesn't work. After all, he's the screenwriter of some of the most intelligent films of the 1970s and 80s – *Butch Cassidy and the*

Sundance Kid, Marathon Man, All the President's Men, The Stepford Wives, The Princess Bride . . .

Goldman has described with real feeling the torment of writing.

'Writing is finally about one thing: going into a room alone and doing it. Putting words on paper that have never been there in quite that way before.' As he says, 'The easiest thing to do on earth is not write.' It's reminiscent of Thomas Mann's definition of a writer – a person for whom writing is more difficult than it is for other people.

So how does Goldman find inspiration for his stories? His screenplay for *Marathon Man* was adapted from his own novel and was made into an iconic thriller staring Dustin Hoffman and Laurence

William Goldman at his writing desk, 1987

Olivier in 1976. Goldman says it was based on two ideas. The first was what would happen if someone in your family wasn't what you thought they were. (That's the Dustin Hoffman character, who thinks his brother is an oil man and actually he's a spy.) As for the second idea: 'I was walking on 47th Street in New York – the diamond district – about forty years ago, and it was a hot day, and all the people that worked in the diamond district were wearing short-sleeve shirts, and you could see all the terrible marks from the concentration camps – because they were all Jewish and they all had their tattoos – and I got the notion: what if the world's most wanted Nazi was walking along this street?'

From these two ideas emerged a compulsive story which climaxes in a torture scene which has put a generation of filmgoers off going to the dentist. The marathon-running student, Dustin Hoffman, has his teeth drilled without anaesthetic by Laurence Olivier, the former Nazis SS dentist at Auschwitz, who repeatedly asks the clueless Hoffman, 'Is it safe?'

Goldman may insist that a walk along New York's 47th Street gave him the germ of his story, but the point is any one of us could have been walking along that street and noticed the tattoos. Only a very few are able to carve a story out of the scene. It's a bit like Michelangelo contemplating his block of marble – his genius is what he takes away, revealing the statue of David inside. Goldman has a mind that makes stories.

If you read through the movie trade magazines, there'll always be a page somewhere advertising a piece of software or a seminar or a course that claims to teach you how to write. You can actually buy an application for your computer that supposedly allows you to build a story. It's as if you can break down a story into a knowable, quantifiable entity that can be proscribed and created according to a formula. William Goldman reckons if it were possible to pre-programme a successful story, we'd all be doing it and making a fortune. Just look at the success of the film *The King's Speech*.

'There's no logic to it. I mean, who in the name of God thinks there's gonna be a successful worldwide movie that wins every honour about a king who has a stammer? It's the worst idea I ever heard but, guess what, it was a fascinating story and it works.'

Casting is surely a vital element in the sense that it can completely alter the way you see the story. *Butch Cassidy and the Sundance Kid* is the most commercially successful of all the films Goldman has written screenplays for. It got him his first Oscar as well as establishing the buddy movie genre. The winning combination of actors Paul Newman and Robert Redford as the eponymous Wild West outlaws was obviously a factor, but Goldman insists that it was a piece of luck. The part of the Kid was supposed to have been played by Steve McQueen, who along with Newman was the biggest movie star in the world at the time. But they couldn't agree on the billing – literally the size and positioning of their names – so McQueen pulled out. Jack Lemon was offered the part but declined because he didn't like horses. Warren Beatty turned it down, as did Marlon Brando, who 'disappeared off with the Indians'. So finally they got Robert Redford. 'But who knows, if we'd had McQueen, if it would have

Butch Cassidy and the Sundance Kid *won Goldman his first Oscar*

been different. I don't know, it might have been better, it might have been worse.'

Very little was known about the real-life characters Butch Cassidy and the Sundance Kid, and Goldman spent years researching their lives before he wrote his screenplay, in which the two men run away to Bolivia.

'When I tried to sell the movie, a guy in a major studio said he would buy it if they didn't go to South America. And I said, "But they went to South America." He said, "I don't give a shit. All I know is that John Wayne don't run away." And I'll never forget that sentence, "John Wayne don't run away."'

Goldman's experience suggests that, although the basic plotlines of stories are universal, our idea of the hero has changed. There was a time when heroes were heroes and, like John Wayne, they didn't run away. What Goldman and his generation did in the late 1960s and early 1970s was to add a new twist – that you could have a hero who is self-deprecating and flawed and not made of granite.

Goldman remembers the thrill of discovering Homer as a child, immersing himself in the tales of the siege of Troy. And then reading Cervantes as a student and flinging his book across the room in a rage when Don Quixote dies. He's almost apologetic in his admiration of these authors.

'I'm gonna say something stupid . . . they were great at story. They really had fabulous stories to tell.'

Ulysses

Irishman James Joyce is acknowledged as one of the great and most original voices in world literature, but to many there's an aura of impenetrability about his work that puts them off. Stephen Fry's Desert Island book on Radio 4 was Joyce's novel *Ulysses* and he urges anybody who's never read it – or tried to and thought that it was too difficult – 'just to let themselves go and swim into it, because, apart from anything else, there were never more beautiful sentences in any single book'.

Ulysses chronicles the journey of the middle-aged Jewish Dubliner Leopold Bloom through the streets of Dublin on an ordinary day, 16 June 1904. To a small but passionate group of devotees, an obsession with the book manifests itself in a rare literary curiosity: Bloomsday. Bloomsday is celebrated every year on 16 June, all over the world, often recreating events from the novel. In Dublin it's frequently experienced as a lengthy pub crawl.

David Norris, a senator in Ireland's parliament, the Taoseaich, is a hugely enthusiastic Joycean and Bloomsday participant. He explains why 16 June was so significant to Joyce and his wife Nora.

'I didn't know Joyce, I didn't know his wife Nora, but I knew all their friends very well and I remember Maria Jolas saying that in the thirties in Paris, when Bloomsday was mentioned, Nora would adopt an insouciant look and she said, "That was the day I made a man out of Jim."'

So why is *Ulysses* lauded so highly? Partly it's the style, the content, and the language, which is a tour de force that has never been matched since, but it's also the humanity of the characters. The genius of the book is that Joyce manages to take examples of Homer's epic adventures of Odysseus (Ulysses in Latin) and find in a single day in Dublin a modern equivalent of the Sirens and the Circe and the Oxen of the Sun. Not only that, but each chapter that represents one of those eighteen adventures has its own colour, atmosphere, smell and linguistic style. The novel contains everything in human life, including public masturbating and use of the c-word that caused it to be banned in the UK until the 1930s. (It was published in Paris in 1922.) It also confronts issues that we now consider to be very urgent and modern, like racism.

The primary beauty of the novel is the hero, Bloom, who is warm, frail, silly and loveable, and yet he's bullied, treated badly, whispered about behind his back. Another wonderful thing about him is that he's so at home with the workings of his own body. The first time you meet him he's having a crap. Joyce describes him shitting and pissing with the same casual exactitude with which he describes him thinking, and it's all treated as part of a continuum. And the character of Bloom, this extraordinary man who we follow over the course of twenty-four hours, grows and grows in stature and warmth and dignity. He becomes a hero out of the most ordinary material imaginable.

David Norris thinks that part of the genius of the book is in its detail. Joyce didn't just invent one day, he investigated and researched that one day in minutest detail. So on that 16 June in 1904, there was a particular horse race that was run, and there was a particular jockey and there was a particular time when the odds on the betting were such and such. And he put that in. He mapped that day completely, perfectly, and Norris believes that's part of the obsession we have with the book to this day. Joyce renders reality with words in a way that a painter can capture the essence of something with brush strokes. *Ulysses* pioneered what became known as the stream-of-consciousness novel, or in Joyce's case, the irrational hiccups of human thinking.

Norris adds: 'Few writers have had more grace and splendour in the way they write. It is just simply beautiful.' He tells the story of

James Joyce, one of the most original voices in world literature

Joyce's great friend, Frank Budgen, who met Joyce in a café in Zurich one day and found him looking rather pleased with himself.

'Good day's work, Joyce?' said Budgen.

'Oh yes', said Joyce.

'Write a chapter, couple of pages, paragraph, a sentence?'

And Joyce replied, 'I had the words and the sentence yesterday but I got the order right today.'

As Joyce said himself in *Finnegans Wake*, it's about getting 'the rite words in the rite order'. But what's probably surprising for those who have been put off reading *Ulysses* because they think it's too difficult or obscure (which is true of his last work, *Finnegans Wake*) is just how much Joyce uses the rhythms and idioms of the street and pub. He had an uncanny ear for the overheard remark.

Norris laughs. 'Every kind of Dublin saying, like "the sock whiskey" for sore legs, for instance, is in it. Joyce collected these things, and I often think that subsequent writers must have thought it terribly unfair competition, because Joyce was so terribly greedy; he left nothing behind for other people.'

He was, to be sure, a hoarder of linguistic treasure.

Davy Bryne's pub is one of the many that grace the pages of the book. Over delicious grilled liver providing an olfactory accompaniment altogether fitting in a novel so keen to that sense, Norris reads out a passage from the opening scene where Bloom is preparing breakfast for his wife Molly.

Mr Leopold Bloom ate with relish, the inner organs of beasts and fowls. He liked thick giblet soup, nutty gizzards, a stuffed roast heart, liverslices fried with crustcrumbs, fried hencods' roes. Most of all he liked grilled mutton kidneys which gave to his palette a fine tang of faintly scented urine.

Kidneys were in his mind as he moved about the kitchen softly, righting her breakfast things on the humpy tray. Gelid light and air were in the kitchen, but out of doors, gentle summer morning everywhere. Made him feel a bit peckish.

The lilt of David Norris' accent adds something to the already lyrical cadence of the writing. Does this Irishness lend to the English language another quality which has helped make Hiberno-English writers so successful? He believes it's because of their discomfort with English. Joyce says, 'My soul frets in the shadow of your language' (meaning the English language). Norris cites a book by Father Peter O'Leary, in which he describes hearing two children talking in the period of the famine. One child says to the other, 'I have no language now, Sheila.' 'Why, what have you got?' she asked him back, and he says, 'I have only English.' And English was, of course, the language of emigration, of administration, of logic and calculation and bureaucracy, whereas Irish was the language of creation, the imagination, improvisation. You find this kind of exuberance in the Irish language, and that cannot be eradicated. So it's as if they're speaking Irish in their mind and translating it into English; and it's that exuberant creative side that is the classic Hiberno-English sound.

Norris adds: 'We tend to be a bit subversive and we're subversive in language too; and people will deliberately use a form of a word that they know is wrong, but the sound of it appeals to them a little bit more.'

Yeats

Declan Kiberd and Barry McGovern are sitting in O'Neills Bar on Suffolk Street, a stone's throw from Trinity College Dublin, where Samuel Beckett studied. Kiberd is Professor of Hiberno-English Literature at neighbouring University College, Joyce's alma mater, and McGovern is an actor, a stalwart of the Abbey Theatre, who created a wildly successful one-man Beckett show. Over pints of Guinness – what else? – the conversation heads into the rich waters of politics and language. Kiberd is a champion of W. B. Yeats.

'I would say the greatness of Yeats was that he took up the oral energies of the people just when the Catholic middle class were trying to transcend them and cast them to one side, and he said no, these are beautiful and they're important. He invented really the modern idea of Irishness and Ireland, and the brilliance of Yeats is that he was both the Irish Shakespeare and its Derek Walcott all rolled into one.'

Kiberd warms to his theme that Yeats was the creator not only of some of the most memorable and patriotic poems, like 'Easter 1916', but also of a cycle of plays that invented Ireland in the same way that Shakespeare's Tudor plays created the idea of England. But he was also postcolonial Ireland's foremost critic, becoming extraordinarily sceptical of and disillusioned with the very country he'd helped to create. Kiberd is convinced that the Bardic idea is probably an essential role for any great writer: you don't just praise the chieftain on good days but you are honest enough to speak the truth and say when things are not being done right.

'And for me that's the magnificence of Yeats, that he was both a tremendous Irish patriot and an absolute auto-critic.'

He believes that Ireland has represented itself through its writers. He tells the story of Oliver Gogarty, a member of the early Irish Senate, who stood up one day and said to the other Senators, 'We wouldn't be here today in a Senate of an independent Ireland, were

W. B. Yeats captured the essence of Irishness

353

it not for the poems of someone like W. B. Yeats.' He was affirming that the word can become incarnate, can become an action.

Easter 1916

I have met them at close of day
Coming with vivid faces
From counter or desk among grey
Eighteenth-century houses.
I have passed with a nod of the head
Or polite meaningless words,
Or have lingered awhile and said
Polite meaningless words,
And thought before I had done
Of a mocking tale or a gibe
To please a companion
Around the fire at the club,
Being certain that they and I
But lived where motley is worn:
All changed, changed utterly:
A terrible beauty is born . . .
I write it out in a verse –
MacDonagh and MacBride
And Connolly and Pearse
Now and in time to be,
Wherever green is worn,
Are changed, changed utterly:
A terrible beauty is born.

It becomes becomes clear in this Dublin pub that an unselfconscious delight in ideas and language are at the heart of good storytelling. There's a great story Yeats picked up in Sligo about a man who went to a cottage and asked for a bed and breakfast, and they said, 'Sure, but you have to tell us a story in return.' And he had no story, so they kicked him out. He went to the next cottage, and they said,

'Yes, you're welcome, but tell us a story.' No story, so he was booted out the door. So he went to a third house and he complained bitterly and described in detail his treatment in the other two houses. So the description of how he was treated for not having a story became the story itself, and he was taken in.

While it's a particularly Irish story, it's also a universal that any culture can understand and appreciate.

'You make your destitution sumptuous, and that's what Samuel Beckett did and why in some ways he is perhaps the central voice of this culture,' Kiberd concludes.

'In the last ditch,' he concludes, quoting Becket, 'all we can do is sing.'

In many ways it was a serendipitous time to be an Irish writer in the late nineteenth and early twentieth centuries – comparable, perhaps, to being an English writer in the late sixteenth and early seventeenth centuries. Just as Ireland created linguistic gold out of the cataclysmic changes after the Hunger, there's something about the language in Elizabethan England that was growing and developing as the country found itself: the King James Bible, the outpouring of plays, the new words that were emerging and a new confidence and boldness in using them. And then, of course, there was Shakespeare.

Shakespeare and the Three Thespians

Imagine you could go back 400 years or so to wander the streets of Elizabethan London. You might pop into a tavern, hoping to bump into Shakespeare or Marlowe or Jonson or Webster, and watch as they drank and talked and scribbled, or go to the newly built Globe Theatre in Southwark on the south bank of the Thames and pay your penny to stand with the other 'groundlings' in the pit in front of the stage and experience for a few hours that extraordinary flowering of language and theatre that the world has never seen since. Shakespeare might be there acting in one of his own plays;

and – it being 1600 – the performance on stage that night might well be his latest work, *Hamlet*.

There was a sort of explosion of words taking place in early modern English in the sixteenth century. What Shakespeare did was to give style and structure to the language, to mine its rich seam and to add to its vocabulary. He invented, or was the first to put in print, around 3,000 new words. You may think you don't know any Shakespeare but of course you do. Expressions like *in one fell swoop* and *it's not the be all and end all*; or *make your hairs stand on end, cruel to be kind, method in his madness, too much of a good thing, in my heart of hearts* and *the long and short of it. Eaten out of house and home, love is blind, foregone conclusion* – the list goes on and on, and they're all creations of Shakespeare. Clichés today, perhaps, but genius.

Someone (Stephen Fry, as it happens) once wrote that as theatre is a rhetorical medium and film is an action medium, so the perfect film hero is Lassie. The hero doesn't need to speak. The boy's on the

The Globe Theatre, c.1600

cliff edge, Lassie looks, Lassie grabs the trouser leg, pulls the boy back . . . you just watch the action unfold. And the perfect theatrical hero is Hamlet, because everything is expressed in language, absolutely everything. It's rhetoric, and he does it like no one else on earth. Hamlet explores sex, life and death, hope, revenge and despair. He's utterly contemporary.

For this reason, the role of Hamlet is seen as the ultimate test for an actor, a theatrical Everest. The character is so full of complexities and the play itself is so well known that sleepless nights are spent worrying about how to bring something new to some of the most quoted lines in literature. A roll call of theatre greats have played Hamlet – John Gielgud, Laurence Olivier, Richard Burton, Peter O'Toole, Derek Jacobi, Ian McKellan. Three actors who've all taken the part of the Prince in the last decade talk about the role: Simon

Laurence Olivier as Hamlet, 1948

357

Russell Beale played the student prince at the National Theatre in 2000 to rave reviews; David Tennant, best known as BBC TV's Dr Who, was Hamlet in a Royal Shakespeare production in 2008; and Mark Rylance has played Hamlet three times – first as a teenager, then with the RSC in 1989 and again at the Globe Theatre in 2000.

'Absolutely terrifying' is how Simon describes performing the 'To be, or not to be' soliloquy for the first time on stage. 'Apart from it being so famous, it was scary because it's such a simple question.' He says it took him until well through the run before he got to grips with the power of the soliloquy – its calmness and self-control.

'I get the sense that it was a radical exploration of a single human soul in a way that hadn't been done before.'

Stephen Fry and Simon Russell Beale in the Globe Theatre

For the Elizabethan audience, *Hamlet* must have seemed incredibly modern and cutting-edge. *Macbeth*, for instance, was set 400 years before, but *Hamlet* is completely different; he has a whole different set of morals and a whole new outlook on the world. The controversial American scholar Harold Bloom wrote a book called *Shakespeare: The Invention of the Human* in which he claimed that before Shakespeare there were no real, rounded, ambiguous, complex characters.

Some people argue the culture we live in, with the competing attractions of television and books and computer games, makes Shakespeare too boring for an awful lot of people, simply not important. Is there any argument Simon can propose that would persuade someone to feel that Shakespeare is worth trying, that he's not an unpleasant medicine that you have to take for the good of your soul?

Simon admits first off that Shakespeare does take work – there's no way round it. He was lucky at school because he had teachers who managed to inspire him. He admits that there are boring bits in Shakespeare – bad bits occasionally – but it's worth the effort: 'It really does yield extraordinary riches.'

The passion which Shakespeare inspires was evident when Simon was touring with *Hamlet* in Eastern Europe. One day he arrived in Belgrade, where the play's poster with his face on it had been stuck up everywhere. He went into a shoe shop, and the woman serving shouted, 'Hamlet! Hamlet!' when she saw him and just kept saying, 'To be or not to be,' over and over again. The idea of that simple little phrase being repeated everywhere they went – even in China – is rather moving.

'Utterly terrifying, poleaxing with fear because of all the baggage that it brings with you and because of the expectation that people have.' That's how David Tennant remembers feeling before his opening night as Hamlet at the National Theatre in 2008. He'd wanted the part since he was an eighteen-year-old drama student and he'd seen Mark Rylance play Hamlet when the RSC came to Glasgow.

'I saw him at that very formative age and . . . it sort of sang. And you suddenly realize that these plays are deeper and wider and longer and better. Utterly contemporary, which is sort of a magic trick because it remains four hundred years old, and yet it seems to keep being reborn and rediscovered.'

Accepting the role of Hamlet – 'like keeping goal for Scotland' – was a brave thing to do. David was at the height of his TV career playing Dr Who and was put under intense media scrutiny in the run-up to opening night. David describes the newspaper articles with critics drawing up their top ten Hamlets of all time and wondering where David would fit in the list.

'And you think, please Lord, let me just remember the lines. And then on the first night the News 24 truck draws up outside your dressing-room window, and you think, oh so now I'm going to fail on a global scale. Terrifying but also so thrilling to have those words at your command and to have that part in your palm for even a brief time. And you think, how do I begin? And of course you just begin

David Tennant holding Yorick's skull, donated by André Tchaikovsky

by not worrying about it, which sounds terribly simple and isn't, but there's sort of no way round it other than just going, "Well, this character happens to say these lines here and they're the first time they've ever been said."'

Richard Burton said it helped him to remember that there's always someone in the audience who will never have heard a single line of *Hamlet* before, so this is absolutely the first time for them. For other audience members, this is not the case. One night during a performance at the Old Vic Burton heard a dull rumble coming from the stalls; it was Winston Churchill sitting in the front row, reciting the words along with him.

Another of David's favourite scenes is Act 2, Scene 2, when Hamlet says to Rosencrantz and Guildenstern, 'I can be bounded in a nutshell and count myself the king of infinite space, were it not that I have bad dreams'.

'You just get the sense that he hasn't slept for days,' says David. 'He wants to sleep but can't. And if you've ever had those long nights of the soul . . . it's just that sense which is so vivid in that speech of "all I want to do is close my eyes but I can't because it's terrifying when I do, because my brain is so full of demons, and I hate myself so much".'

Shakespeare talked a lot about sleep. He probably didn't get enough of it or maybe he simply loved to sleep. The language he used – in *Julius Caesar* he calls it the 'honey-heavy dew of slumber'; in *Macbeth* it's 'sore labour's bath . . . balm of hurt minds' – isn't that fabulous?

The most identifiably visual moment of the play is the gravediggers' scene, when Hamlet holds up the skull of the court jester and says, 'Alas poor Yorick, I knew him'. It's a memento of death, just like the 'to be' soliloquy. 'I think the Yorick moment is much more specific,' says David 'because he's looking at the material of a human being and he's imagining that lips hung here and he's trying to get his head round the actuality of death.'

The skull used in David's performance was in fact a real one, donated by a concert pianist called André Tchaikovsky, who donated his head to the RSC in his will, to be used as a Yorick.

'The first few performances holding a real human head was terribly potent because that is exactly what it's about – we will become inanimate matter.'

As a young boy Mark Rylance spoke too fast to be understood by anyone. He had elocution lessons to slow him down, chanting tongue twisters and reciting poems and prose out loud. Mark found that learning bits of Shakespeare by heart and speaking the lines in front of people was the first time that he was able to express a whole range of emotions and ideas. He performed his first Hamlet as a sixteen-year-old school boy, then played him again aged twenty-eight and finally at forty. He reckons that's about 400 performances in all.

When Mark was offered the part of Hamlet at the Royal Shakespeare Company it was a huge thing for him. He told the director, Peter Gill, that he planned to go up to Stratford immediately and read the old prompt scripts of all the luminaries who'd played the part, 'like it was some big oak tree and I hoped I might add a little twig to the tree by being aware of all the choices they'd made'. Peter Gill told him not to be a fool – they're all dead and gone or at least not playing the part any more. '"It's you who are alive now. Make sure it's not set in outer space, but apart from that it's you." I said, "Yeah, but David Warner, he has such a . . ." "He was only wonderful because he was of his time," said Gill, "and you're of our time." And that comment comes right down to the last ten seconds before you're about to enter to say "To be, or not to be".'

Just like Richard Burton, Mark had encounters with members of the audience who knew the play a bit too well. He had to be careful not to make his dramatic pauses too long. 'In Pittsburgh there was a little old lady sitting next to her husband, and I came out right next to her in my pyjamas, all tearful and crying, and I said, "To be, or not to be". And then I thought for a moment. And she turns to her husband and says, "That is the question". And everybody heard it and laughed a bit. But I was able to say, "That *is* the question!"'

Those sort of close-hand experiences with the audience stood Mark in good stead when he was appointed the first artistic director of the newly recreated Globe Theatre in 1995. He developed strong ideas about the relationship between actors and the audience, so that by the end of his ten years there he says he thought of the audience as more like fellow players. They were bringing the most important energy of the whole evening, a desire to hear a play. Mark likens it to a moment when he was playing Hamlet and delivering the line 'Sit still, my soul: foul deeds will rise, though all the earth o'erwhelm them, to men's eyes'.

Mark Rylance
as Hamlet

'I remember performing the play out at Broadmoor special hospital and turning to a man and I wasn't aware whether he was a nurse or a patient but I could see his imagination was completely with me. And in that moment between us I felt: who's to say you're the audience and I'm the actor? We're in this together.'

Hamlet's soliloquy

To be, or not to be, that is the question:
Whether 'tis nobler in the mind to suffer
The slings and arrows of outrageous fortune,
Or to take arms against a sea of troubles,
And by opposing end them? To die, to sleep,
No more; and by a sleep to say we end
The heart-ache, and the thousand natural shocks
That flesh is heir to: 'tis a consummation
Devoutly to be wished. To die, to sleep;
To sleep, perchance to dream – ay, there's the rub:
For in that sleep of death what dreams may come,
When we have shuffled off this mortal coil,
Must give us pause – there's the respect
That makes calamity of so long life.
For who would bear the whips and scorns of time,
The oppressor's wrong, the proud man's contumely,
The pangs of despised love, the law's delay,
The insolence of office, and the spurns
That patient merit of the unworthy takes,
When he himself might his quietus make
With a bare bodkin? Who would fardels bear,
To grunt and sweat under a weary life,
But that the dread of something after death,
The undiscovered country from whose bourn
No traveller returns, puzzles the will,
And makes us rather bear those ills we have
Than fly to others that we know not of?
Thus conscience does make cowards of us all,
And thus the native hue of resolution
Is sicklied o'er with the pale cast of thought,
And enterprises of great pitch and moment,
With this regard their currents turn awry,
And lose the name of action.

The Professor and Bob Dylan

Sir Christopher Ricks is one of the greatest literary critics of our day. He's written seminal works on Milton, Keats, Tennyson, Beckett, T. S. Eliot, edited the *Oxford Book of Verse*, has been Warren Professor at Boston University since 1986 and until 2009 was Professor of Poetry at Oxford University. He's been described as holding all of English poetry in his head. He was also my very exacting professor at Bristol in the 1970s.

Photos of Ricks' academic subjects adorn the walls of his elegantly spacious office at Boston's Editorial Institute, which he now runs – the haggard, intense face of Samuel Beckett, T. S. Eliot besuited and respectable, and Bob Dylan, whom he has favoured with a compelling 500-page critical work entitled *Dylan's Visions of Sin*. Sir Christopher is the critics' critic. Scholarly, yes, with an extraordinary forensic ability for close textual analysis down to the use of the comma or apostrophe, but also very funny and in his seventy-seventh year as bright and acute as ever in not letting any lazy thinking get in the way of good criticism.

Stephen Fry almost had Ricks as his supervisor when they were both at Cambridge University in the 1980s. Now, thirty years later, he finally gets the chance of a one-to-one tutorial with the Professor. The ostensible subject is Beckett and Dylan and why poetry matters, but in typical Ricksian fashion they end up dancing all over the place. Here's a flavour of the Ricks Masterclass, kicking off with the difference between poetry and prose.

CR: I think poetry is to be distinguished always from prose. They have different systems of punctuation.

SF: That's a really good point because it's a game people play to try and devise the best, most compact, most necessary and sufficient definition of poetry as opposed to prose. And you said different punctuation which sounds trivial but it strikes at the heart of it in a way, doesn't it?

Bob Dylan, a poet songwriter

CR: Well, I think that it does. The line endings are significant in poetry or verse. They carry significance.

SF: So it's the shape on the page. T.S. Eliot said that poetry is not about expressing emotion or personality, it's almost the opposite. And people might say, but surely poetry is the excess, the demonstration of personality and emotion?

CR: You're right, especially in a world in which too much is being made of the idea that poetry is self-expression. Though Eliot is always resourceful enough to know that you need a multiplicity of ways of talking about things. That is, a poem is in some ways like a person. In another way it's like a plant. In another way it's like a beautiful building. We need all these figures of speech. The key term from him, I think, is when talking about intelligence either in criticism or in poetry – he thinks of it as judging. It's the discernment of exactly what and how much you feel in any given situation. So it doesn't make poetry or literature the realm of feeling as against prose as the realm of fact or proof or argumentation.

I think it's a very, very beautiful formula in what it does with both thinking and feeling.

SF: And it requires a huge amount of honesty. Direct confrontation.

CR: Great honesty, including doubts about the rhetoric of honesty. So as one says, 'I'll be completely honest with you,' as though in the ordinary way I was doing no such thing.

SF: Yes quite. We've started with Eliot, which is for some people the start of modernism and the time when a lot of the public turned off poetry. But while we're on the subject of him there's a precision, an almost miraculous ability to create lines which stick in the head like music. Anybody who's read Eliot will probably say he's one of the easiest poets to memorize a phrase from. And I wonder where that comes from. Why, for example, 'the young man carbuncular' and not 'the carbuncular young man' in a modern poem?

CR: Well it would partly be that – the striking turn of phrase, and that is indeed a turn, isn't it? What it's doing is imitating languages, ancient or modern, in which an adjective comes after a noun. The human form divine becomes the young man carbuncular and so on. So there's something about moving. You've got to allow a shape to it. So I think he's always interested in resisting either the tyranny of the eye or the tyranny of the ear. Each is inclined to take over. The eye will say every time you use the word 'image' it's clear that imagination is seeing things. But every time you turn to a figure of speech that comes from hearing, you seem to be deaf to this.

SF: That's a very good example of the close attention to language that people might find astonishing in your work. You go almost to the molecular and atomic level of a sentence and into the syllables and the actual structure of words and you find in them an energy and they are often the thing that causes the whole work to show itself.

And I suppose as much as anything, the problem that poets have to confront is that they're making their art out of something that is common to all humanity. Unlike a painter who can go to a shop and get turpentine and acrylics and canvas and brushes or a musician who has a special language of augmented sevenths and special machines made of brass and wood. There's another Eliot phrase – 'to purify the dialect of the tribe'. So a poet has to take the same thing I use when I order a pizza over the phone and turn that into art. And does that mean poetry has two choices? One is to embrace the everyday and the other to try and get rid of it and to find a noble and elevated language.

CR: I think it's always having to do both. I think poetry is very like ordinary life. It's continually wanting some arrangement of things that are new and surprising to it. It's continually interested in the fulfilment of expectations and in the arrival of surprise.

SF: Yes. You want to be surprised but you also want the comfort of reassurance that you know them.

CR: The poet Donald Davy talks about how the words of a poem should succeed one another like the events of a well-told story. They should be at once surprising and just. That is, it's easy to be surprised if you don't care whether the effect is a just one. It's easy to be just if you don't care whether it's surprising. Bob Dylan loves rhyming 'new' and 'true' because every artist is in the business of finding something new to say that is also true. It's easy to find something new to say. It's easy to find something true to say but these people come up with very extraordinary things that are at once new and true.

SF: Again it suggests the Eliot line in the 'Four Quartets' about arriving at a place and seeing it for the first time. There is this sense in poetry and even in just great writing of being in a familiar place and being assured by the authority of a writer that you trust them and yet also being surprised by them because they make you look with new eyes and things. And is that something that you think is innate in that there is a certain class of person who can do this and they do it with words when someone else might do it with music?

CR: Dr Johnson believed that there were talents and you employed them in any way that struck you. It's as if you were at the North Pole and you could walk south in many different directions. But what you do is essentially exactly the same, you walk south. On the other hand some people are amazingly numerate. They can look at a spreadsheet and see that the figures have been rigged and fixed. They can see that the books are being cooked. I could look at equations and symbols and numbers for ever and be blank. What about you?

SF: I'm exactly the same. I don't understand. And as you say some people are like a spider on the web – every twitch of the filament means something to them. And they can chase it down. So yes there is the gift of language.

CR: Well, I think poets are really intelligent, and resourceful poets are much less vulnerable than you might think because their self criticism is alive to what might be the criticism of others. Pretty well all great writing has a warning about over-valuing writing in it. At some point or other Shakespeare will tell you not that plays or dramatic representations tell no truths but that you must remember they also tell lies. And so words half reveal and half conceal the truth within. And every great writer has intimated some such things at some point.

SF: I think that's absolutely right. Language is a bit like a dress. A dress can reveal the form, can flatter the form, can exaggerate the form, can give a real sense of the beauty and elegance of a particular form but it also hides deformities. It also masks and covers nudity; covers the passionate side of us, our fleshly side. And words are constantly doing that.

There's also this idea that there's great literature: Leavis and his great tradition and Harold Bloom and his sense of the canon of writers, but can they include in that twentieth- and twenty-first-century figures from what is often called popular culture – like Bob Dylan who you've written about?

CR: I think Dylan uses words with extraordinary effect. The effect is related to a different system of punctuation – the speed and pace at which it goes is determined by him, his music and his voice. He changes those and the beautiful thing he says about the songs is, my songs lead their own lives, and it's lives in the plural not because each song has a life but because each song has lots of lives. I think again and again Dylan is very good when you can imagine an unimaginative creative writing school telling him he'd got it wrong. I think that he's simply astonishingly imaginative with words.

Funeral Blues

Stop all the clocks, cut off the telephone,
Prevent the dog from barking with a juicy bone,
Silence the pianos and with muffled drum
Bring out the coffin, let the mourners come.

Let aeroplanes circle moaning overhead
Scribbling on the sky the message He is Dead.
Put crepe bows round the white necks of the public
   doves,
Let the traffic policemen wear black cotton gloves.

He was my North, my South, my East and West,
My working week and my Sunday rest,
My noon, my midnight, my talk, my song;
I thought that love would last forever: I was wrong.

The stars are not wanted now; put out every one,
Pack up the moon and dismantle the sun,
Pour away the ocean and sweep up the wood;
For nothing now can ever come to any good.

That poem was written by W. H. Auden, but you may well know it better from the film *Four Weddings and a Funeral*, where it was recited during the funeral of the title. It's extraordinary how something can have such impact, be so succinct and have such emotional truth behind it.

For anyone who's had to organize a funeral, the choice of words is one of the most difficult things to get right. In the past, the traditional funeral passages from the King James Bible had little competition. But in our more secular times many prefer not to invoke religion at all. So what do we do when we want to express the grief and love and sadness of the loss of someone? Music is integral, and, if it's right, the emotional heft will inevitably lead to tears.

So what led *Four Weddings* scriptwriter Richard Curtis to choose the Auden poem for the funeral scene, a choice which catapulted sales of Auden's poetry beyond anything he'd enjoyed while still alive?

Richard explains with his unfailing modesty that he didn't feel up to the job.

'Tragically in my life, in every film I've ever done, the single best moment in the film has nothing to do with me at all. I was writing a moving funeral scene so I thought I'd better leave it to a better man. I'd always been told I should study Auden and I didn't understand most of his poems. And I remember being very thrilled when I came across that one. And I think it's no coincidence that it's in fact called 'Funeral Blues' and it's a lyric; it was meant to be sung. And that is symptomatic of the fact that I'm passionate about lyrics in a way more than poems.'

It's become *the* thing to choose songs rather hymns or prayers at funerals. 'I Did It My Way', 'Je Ne Regrette Rien' or 'Angels' may be a bit clichéd, but people clearly feel their lyrics do the job better than a poem or a reading. They can express a communal emotion that everybody can share.

Richard has a theory that we don't have access to poems now in the way we once did. The Romantic poets were celebrities in their day. People were outraged by the work of Byron because they knew about him, he was famous. Nowadays it's hard for a poem to break through into the popular culture, so what happened with *Four Weddings* was a rare example of a poem being heard by enough people to get a passionate reaction.

Poems are often perfect word for word, pop lyrics less so, although, as Richard is keen to point out, there are some wonderful wordsmiths in the world of pop lyrics. He cites Paul Simon as one lyricist who has written some extraordinary, powerful songs.

'Every day I think of that line from "The Boxer": "A man hears what he wants to hear and disregards the rest." And it lodges in your head, and as you go through life you realize people are only hearing a bit of what you say because it's the bit that suits them. Rufus Wainwright's song "Dinner At Eight" is about him and his father, and there is no finer expression of the argument between a son and a father who abandoned him. And there are the huge, big popular ones that everybody knows. They become the fabric of your life, and those lyrics are carried around with you and reflect your moods and feelings.'

Richard is in full flow now. 'If you pick up a poem for the first time you have to piece it together. It's much harder. Whereas a song by Coldplay like "Fix You" has the lyrics "I will try to fix you." It's very direct, and the fact that the lyrics may not be as well crafted is compensated by the beauty of the tune, and is enough to turn it into something deeper. And on top of that you have the feeling that your whole generation heard that song together, so it has a binding effect. If you stood in a stadium with 45,000 other people who know those words, it's the Nuremberg Rally of pop.'

He's right, of course. Pop songs are a brilliant way of people sharing a culture. But, despite his love of popular lyrics, does Richard still read poetry? He had to read lots of it at Oxford. He laughs.

'I think there was a six-month period in which I understood it. I tried to read something by Yeats the other day which I know used to be my favourite poem. It's gone completely. It's as if I've forgotten the language.'

Paul Simon, right, with Art Garfunkel

Music

If you've ever wondered why you sing in the bath (if you don't, ignore this bit); or why you spent all those hours and hours as a teenager, shutting out the annoying world of parents and other people, glued to Radio 1 and the charts; or why the hairs on your neck prickle and your heart beats faster whenever Wagner's music fills the air – well, if you've ever wondered, sorry, but nobody really knows. To be precise, scientists can't agree on what it is about music and singing that touches us so profoundly, and whether it's something to do with our evolution. Are we hard-wired to be musical?

Music is universal. It's found in all cultures across the ages, and archaeologists have unearthed musical instruments dating from as far back as 34,000 BC. The mystery is *why* humans have been singing and making music virtually since prehistoric times. What purpose does it serve?

Evolutionary psychologists have a few theories. Some think that music originated as a way for males to impress and attract females, rather like brightly coloured birds competing with each other to produce the most elaborate and complex songs. In other words, it's a tool of evolution and natural selection: the male with the biggest lungs and catchiest song gets the girl. Another idea is that women were the original music-makers, and it goes all the way back to the universal instinct of the mother crooning and singing to her child. Experiments show that mothers automatically make their speech more musical when they talk to their babies, more lilting and melodic, and the theory is that music perhaps evolved as a sort of prehistoric baby-pacifying tool. Human babies can't just cling on to their mothers' bodies the way other primates do, so perhaps the singing was a way of the mother keeping contact with her children when she had to put them down to work.

And then there's a third theory that identifies music as a sort of social glue, a way of bonding early human communities, much in the same way that football supporters or people in church or families round the piano singing together enhances a sense of tribal identity. The evolutionary psychologists trace this back to the necessity for early tribes to work together for survival: communal singing demands coordination, bringing many voices together, and the theory goes that this is a way of practising for the kind of teamwork crucial in hunting or fighting for survival.

Interestingly, researchers at the Montreal Neurological Institute discovered, when they scanned musicians' brains, that listening to music stimulated exactly the same part of the brain which food and sex affect. In other words, music lit up the basic, instinctual, pleasure centres of the brain.

However, others dismiss these ideas. The Harvard linguist Steven Pinker caused an uproar when he addressed a conference of cognitive psychologists in 1997 and told them that their field of music perception was, basically, a waste of time because music is just an evolutionary accident, a redundant by-product of language: 'Music is auditory cheesecake,' he said, 'an exquisite confection crafted to tickle the sensitive spots of several of our mental faculties.'

For many people – whatever the neurological and evolutionary theories – music is another form of language. A way of communicating without words.

Music is universal and serves as a social glue

Much research has been done on the use of music therapy with people suffering from dementia. They find verbal communication very difficult because the disease causes aphasia and amnesia. If you can't find the words, if you can't remember who you are, how do you express yourself and connect yourself to the people and the frightening world around you? Music therapists use singing and music *in place of* words. Singing a song with someone with Alzheimer's doesn't put demands on them; it doesn't require the answer to a question; it doesn't make the world even more confusing than it is. Singing is a way of being together, of inviting the person to take part, of somehow bypassing the damaged part of the brain that can no longer form the words. It releases tension and calms anxiety because it opens a door to expression when all the other doors are barred and shut tight. Music is part of who we are. It travels further, down and down into that part of ourselves which is older, deeper, mysterious.

Auditory cheesecake indeed.

Advertising

If you were to chant 'Helps you work, rest and play', the chances are most people will respond with 'A Mars a day'. Or if you sing, 'Now hands that do dishes . . .', a surprising number of you will feel the urge to sing 'with mild green Fairy Liquid'. Someone younger will have the same automatic response to 'Just do it', immediately associating it with Nike. For just as succinct language can have a powerful effect on us through poetry and song, so the perfectly turned phrase can enter our subconscious, influencing our actions and decision-making. Nowhere is the use of the clever slogan seen more explicitly than in the language of advertising.

From an ancient Egyptian town crier hired to shout out news of the arrival of a goods ship to a computer pop-up ad offering the secrets to a flat belly, we have always found ways of attracting the public's attention to a product or business. Advertising may have become much more sophisticated, but wherever communities and commerce exist, so too does the advert in some shape or another.

We're been shouting our wares and promoting our products for thousands of years. Historians reckon that outdoor shop signs were civilization's first adverts. Five thousand years ago the Babylonians hung the symbols of their trades over their shop doors, a practice still used today in areas of poor literacy or in some of the traditional shops like barbers – the red-and-white pole – or the three golden balls of the pawnbroker. A poster found in Thebes from 1000 BC offers a gold coin for the capture of a runaway slave. The Romans advertised the latest gladiator fight on papyrus posters and seem to have introduced the world's first billboards with their practice of whitewashing walls and painting announcements on them.

The advent of printing and spread of literacy meant advertising could expand into handbills and newspapers. A newspaper ad for toothpaste appeared in the *London Gazette* in 1660: 'Most excellent and proved Dentifrice to scour and cleanse the Teeth, making them white as ivory.'

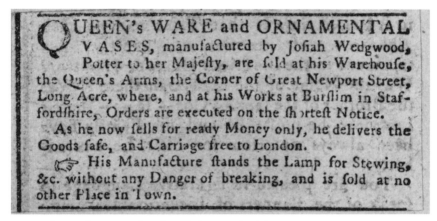

An early example of a Wedgwood ad in a newspaper, c.1769

The Industrial Revolution and the greater availability of factory-produced consumer goods brought a growing awareness amongst manufacturers that they could *create* a need for a product amongst the middle and lower classes able to afford luxury items. Our friend Dr Samuel Johnson wrote one of the first articles on advertising in a 1759 edition of his magazine *The Idler*: 'Promise, large Promise, is the soul of an Advertisement . . . The trade of advertising is now so near to perfection, that it is not easy to propose any improvement.' He wasn't very prescient, for at that very moment one Joshua Wedgwood was founding his pottery workshop and he had very big ambitions about improving advertising technique. Wedgwood was one of the first industrialists to recognize the importance of creating a market through advertising and used newspapers ads, posters, publicity stunts, give-away promotions and money-back guarantees to persuade people of the need for a must-have Wedgwood vase or a twenty-piece dinner service.

Increased competition amongst manufacturers and a growing public sophistication opened the way for agencies in the late nineteenth century who promised to create and run advertising campaigns for the client. Advertising had become a profession. The advent of radio and television extended the mass reach of advertising, revolutionizing its persuasive potential. This is the world of advertising we're all familiar with – the TV commercials and product placements in films and internet pop-ups.

On the reception desk of the west London ad agency Leo Burnett is a large bowl of green apples, a reminder of the humble beginnings of the founder, Mr Burnett, who established his agency in Chicago in 1935 with just one account, a staff of eight and a bowl of apples on the front desk. Legend goes that, when word got around Depression-hit Chicago that Leo Burnett was giving apples to visitors, a newspaper columnist wrote, 'It won't be long till Leo Burnett is selling apples on the street corner instead of giving them away.' Leo was a wizard with visual imagery and created the iconic brands for the Jolly Green Giant canned peas and corn, Kellogg's Tony the Tiger – 'They're GR-R-R-E-A-T' – and most famously the Marlboro Man for Philip Morris, who were persuaded to repackage their woman's cigarette as a rugged man's smoke. Today, Leo Burnett is one of the world's leading advertising organizations.

Don Bowen is a creative director at the agency's London office and currently in charge of the Kellogg's and Daz accounts. He began in the business thirty years ago as a copywriter and manages to turn on its head the adage of a picture being worth a thousand words.

'Words are tremendously important in advertising because there have been hardly any ads where there have been only images that can make a lot of sense. *The Economist* has done this, once or twice, but most of the time most ads have words in them, and very often the words can be worth a thousand pictures.'

Don gives the example of the Volkswagen advertising campaign in the 1960s. The American ad agency DDB faced an apparently impossible task – selling Hitler's favourite car to the Americans, a decade or so after the Second World War. What they came up with

Lemon.

This Volkswagen missed the boat.

The chrome strip on the glove compartment is blemished and must be replaced. Chances are you wouldn't have noticed it; Inspector Kurt Kroner did.

There are 3,389 men at our Wolfsburg factory with only one job: to inspect Volkswagens at each stage of production. (3000 Volkswagens are produced daily; there are more inspectors than cars.)

Every shock absorber is tested (spot checking won't do), every windshield is scanned. VWs have been rejected for surface scratches barely visible to the eye.

Final inspection is really something! VW inspectors run each car off the line onto the Funktionsprüfstand (car test stand), tote up 189 check points, gun ahead to the automatic brake stand, and say "no" to one VW out of fifty.

This preoccupation with detail means the VW lasts longer and requires less maintenance, by and large, than other cars. (It also means a used VW depreciates less than any other car.)

We pluck the lemons; you get the plums.

The 1960s Volkswagon campaign was part of a creative revolution in advertising is known simply as 'The Lemon' – a black-and-white picture of a Beetle car with the one word, 'Lemon', underneath. The story below explained that this car had been rejected by the quality-control inspectors because of a blemish on the glove box. 'We pluck the lemon; you get the plums,' ran the end tag. It intrigued readers that a company should be critical of one of its own cars. Then it hooked them with its assurances of rigorous inspection. It was an approach to advertising that the Lemon's creator, William Bernbach, saw as part of a creative revolution, rethinking how advertising works. (For the car rental company Avis, his agency made an apparent failure a virtue: 'We're only Number 2. We try harder.')

'Lemon' is an example of what copywriters call the endline, like 'Probably the best lager in the world,' or 'Refreshes the parts other beers cannot reach,' or 'Don't just book it, Thomas Cook it.' Huge amounts of time are spent getting them right, distilling a number of ideas into one three-, four- or five-word end line. 'English is a particularly good language for being able to play gags, to word play, to sum up an idea pithily,' says Bowen, 'like "You'll never put a better bit of butter on your knife."'

TV commercials in the UK particularly like to use endlines and some of the best slogans (and quirky storylines) have become as much a part of our culture as classic TV shows. Many of us can recite advert slogans from our childhood as easily as we can chant a nursery rhyme. 'Don't forget the Fruit Gums, Mum,' Beanz Means Heinz, 'All because the lady loves Milk Tray.' 'Opal Fruits! Made to make your mouth water', 'For Mash get Smash!', 'Happiness is a cigar called Hamlet'

'The end line has got to resonate with people,' says Bowen. 'There are thousands of end lines out there, most of which we don't care about. But when you get it right and there's some emotional connection with the end line, then I think you can tell the story which leads up to it again and again, which is what makes the words so powerful.'

Today the notion of advertising as an art of persuasion is seen as rather an old model. The days of the unique selling propositions – USPs – have gone. One detergent really isn't that different from another. So you have to attract people to your brand rather than persuade them that it's the right product. Advertising has become much more about engaging people emotionally, and the most successful way of doing this is by creating a story about the brand.

Don's favourite ad from his own agency in recent years is the McDonald's poems.

> *Now the laborers and cablers and council-motion tablers*
> *were just passing by . . .*
> *and the first-in types and lurking types and like-to lose*
> *their-gherkin types and suddenly-just-burst-in types*
> *were just passing by.*

He likes the attention to detail of the words, the internal rhymes, the little word jokes. The most interesting thing of all was – 'it was so human'.

Does Bowen worry that this storytelling in advertising, the communal appeal of the sales pitch, will be lost as the consumer is targeted with individual messages and offers via the internet?

'I think there will always be good stories to tell. Whenever it's difficult in advertising, I always say to myself, this brand, with any luck, is going to be around in fifty years, and somebody is going to need to advertise it. So there will be a story to tell. The question is just finding it.'

The Most Famous Ad Man in the World

He has the looks, the pipe and the suit of one of the original Mad Men in the American TV drama set in the advertising world of the 1950s and 60s. Dubbed the 'King of Madison Avenue' and the 'Pope of Modern Advertising', David Ogilvy rewrote the rules of advertising with his emphasis on brand image, consumer research and the Big Idea. 'Unless your advertising contains a big idea,' he insisted, 'it will pass like a ship in the night.'

Born in Surrey in 1911, this eccentric Anglo-Scot started his working life as a sous-chef in Paris after being sent down from Oxford University. A job selling Agas, the most expensive

Iconic adverts demonstrate the power of the word

cookers on the market, door to door in Scotland during the Depression gave him the experience of direct selling (albeit to stately homes and convents) which influenced his later career. He wrote the company's sales manual, *The Theory and Practice of Selling the Aga Cooker*, which began 'In Great Britain, there are twelve million households. One million of these own motor cars. Only ten thousand own Aga Cookers. No household which can afford a motor car can afford to be without an Aga.' The twenty-four-year-old Ogilvy offered a variety of personal tips including: 'The good salesman combines the tenacity of a bull dog with the manners of a spaniel. If you have any charm, ooze it.'

After some initial training in the ad agency Mather & Crowther, he left London at the end of the 1930s to seek his fortune in the United States. There he was hired by pollster George Gallup to work with him at his newly founded audience research institute; he and Ogilvy spent months travelling from New Jersey to Hollywood and back, quizzing the public about their movie star preferences and selling the research to studio heads and movie producers. After working for the British Secret Intelligence during the war and a brief period living with the Amish as a tobacco farmer, Ogilvy finally moved to New York

David Ogilvy and Mad Men's *Don Draper*

in 1948. With virtually no experience in advertising and, as he recalled, 'no credentials, no clients and only $6,000 in the bank', he opened up his own agency in a two-roomed office with a staff of two. Within a decade he'd built Ogilvy & Mather into an advertising powerhouse that attracted the biggest clients in America. He employed all the skills he'd learned in door-to-door selling and audience research along with a flair for language, a strong visual sense and a personal showmanship – he was known to dress in a full-length black cape with scarlet lining.

The bulk of Ogilvy's notable advertising campaigns were produced in these early years. He called them his Big Ideas. He outlined how to recognize one by asking five questions. 'Did it make me gasp when I first saw it? Do I wish I had thought of it myself? Is it unique? Does it fit the strategy to perfection? Could it be used for thirty years?'

An early iconic campaign was for a small clothing company called Hathaway. On the way to the photo shoot, Ogilvy stopped to buy some eye patches. The photograph, 'The Man in the Hathaway shirt', caused a sensation. The clothing firm couldn't keep up with demand. The eye patch had given the ad what Ogilvy called 'Story Appeal'. The reader was intrigued by the model with one eye and wanted to find out more. Other campaign successes followed: 'At sixty miles an hour the loudest noise in this new Rolls-Royce comes from the electric clock,' and 'Only Dove is one-quarter cleansing cream' (which helped make Dove America's bestselling soap). When Ogilvy won the Shell account in 1960, other major American advertisers – General Foods, Campbell Soups, American Express – jumped on board.

Ogilvy was credited with introducing the novel idea (for the 1950s certainly) of the intelligent consumer. In his bestselling *Confessions of an Advertising Man* he included his most famous aphorism 'The consumer is not a moron. She is your wife. Don't insult her.'

Words were the backbone of Ogilvy's campaigns. His biographer, Kenneth Roman, worked for him for twenty-six years and described the ad man at work.

The man in the Hathaway shirt

AMERICAN MEN are beginning to realize that it is ridiculous to buy good suits and then spoil the effect by wearing an ordinary, mass-produced shirt. Hence the growing popularity of HATHAWAY shirts, which are in a class by themselves.

HATHAWAY shirts *wear* infinitely longer—a matter of years. They make you look younger and more distinguished, because of the subtle way HATHAWAY cut collars. The whole shirt is tailored more *generously*, and is therefore more *comfortable*. The tails are longer, and stay in your trousers. The buttons are mother-of-pearl. Even the stitching has an ante-bellum elegance about it.

Above all, HATHAWAY make their shirts of remarkable *fabrics*, collected from the four corners of the earth—Viyella and Aertex from England, woolen taffeta from Scotland, Sea Island cotton from the West Indies, hand-woven madras from India, broadcloth from Manchester, linen batiste from Paris, hand-blocked silks from England, exclusive cottons from the best weavers in America. You will get a great deal of quiet satisfaction out of wearing shirts which are in such impeccable taste.

HATHAWAY shirts are made by a small company of dedicated craftsmen in the little town of Waterville, Maine. They have been at it, man and boy, for one hundred and fifteen years.

At better stores everywhere, or write C. F. HATHAWAY, Waterville, Maine, for the name of your nearest store. In New York, telephone MU 9-4157. Prices from $5.50 to $25.00.

The eye patch gave the ad 'Story Appeal' and caused a sensation

386

> *Being edited by Ogilvy was like being operated on by a*
> *great surgeon who could put his hand on the only tender*
> *organ in your body. You could feel him put his finger on*
> *the wrong word, the soft phrase, the incomplete thought.*
> *But there was no pride of authorship, and he could be quite*
> *self-critical. Someone found a personally notated copy of*
> *one of his books in which he had written cross comments*
> *about his own writing: 'Rubbish,' 'Rot!' 'Nonsense.' He*
> *would send his major documents around for comment,*
> *with a note: 'Please improve.'*

Ogilvy retreated to a chateau in France in the 1970s, unconvinced by the direction advertising was taking – less direct-sell, more art form.

> *I do not regard advertising as entertainment or art form,*
> *but as a medium of information. When I write an*
> *advertisement, I don't want you to tell me that you find it*
> *'creative'. I want you to find it so interesting that you buy*
> *the product. When Aeschines spoke, they said, 'How well*
> *he speaks.' But when Demosthenes spoke, they said, 'Let*
> *us march against Philip.'*

When Ogilvy died in 1999, the Leo Burnet agency placed a full-page ad in the trade papers. It read: 'David Ogilvy 1911 – Great brands live forever.'

Oratory

Oratory is a powerful ally of the written word because a great public speech is a potent means of communication. It can persuade, move, convince, agitate, enlighten and – yes – manipulate, and history is littered with examples of oratorical tours de force, from Cicero's attacks on Mark Antony to Martin Luther King's 'I have a dream'.

The art of oratory goes back to ancient Greece, where public speaking was considered an essential part of education. Socrates,

Aristotle and Plato discuss it at length, and it remained a central part of Western liberal humanistic education into the twentieth century. Lawyers, politicians and entertainers all need to be good orators, and it comes in handy if you're asked to be a best man or after-dinner speaker. Wit, humour and the habit of reasoned arguments are all part of the armoury of the orator. It's not essential but it helps if you believe in what you're saying, as the best examples of oratory bear out.

The art of oratory is more than just the words and the arguments. The American writer Gore Vidal pointed out that in ancient Rome the senators were really drama critics, critiquing not only the contents of one another's speeches, but the style of delivery. And former President Bill Clinton says: 'A lot of communication has nothing to do with the words; a lot of it is just your body language, or your tone of voice, or the way you look in your eyes'.

Clinton is one of the best-placed (and highest-paid) contemporary orators to remind us of the challenge in delivering a truly great speech:

'You measure the impact of your words,' he says, 'not on the beauty or the emotion of the moment but on whether you change not only the way people think, but the way they feel.'

There is fine oratory to be found in fiction too. Shakespeare is a great place to look, of course: Mark Antony's funeral oration over the body of Julius Caesar in Act 3. The opening words – 'Friends, Romans, countrymen, lend me your ears!' – along with 'To be, or not to be' and 'Out, damn spot' and a dozen others – must be among the most-often-quoted lines of Shakespeare, but the whole speech is a massive 137 lines of verse, in which Antony works the crowd and succeeds in changing the minds and hearts of the Roman mob. He uses every oratorical trick in the book, every tool of rhetoric to turn their hostility towards him and Caesar into grief for their murdered emperor and rage against Brutus. Gradually, he builds his argument by suggestion, ambiguity, calculated flashes of anger and grief, hypnotically repeated cadences – 'Brutus is an honourable man' – and by persuasion, artifice, sneaky cleverness and superb rhetoric, he brings the crowd over to his side. It's a piece of theatre, performed by a duplicitous man who is a master of oratory.

Back to real orators, past and present. There are all kinds of different styles of oratory, and obviously they change over time as tastes and traditions change. The demagogic style of someone like Hitler certainly did work once, but we're not taken in by it in the same way any more. The Reverend Jesse Jackson is one of the most powerful speakers to have come from a background of preaching, a tradition which creates a kind of poetry of words and rhythm. 'Words,' he says, 'paint pictures; words draw our imagination', and help us to believe we can achieve great things.

The Reverend Martin Luther King's famous 'I have a dream' speech in 1963, and Obama's 'Yes, we can' drew heavily on this tradition of church preaching. The pastor is both the composer and the conductor, delivering the words and orchestrating the response from the audience by the pace of delivery, the rhythm of

The Reverend Jesse Jackson, one of the most powerful preachers in America

the words and pauses, the rhetorical questions, the 'call and response' sections where he knows the audience will murmur in agreement, or repeat a phrase. You can hear it when you listen to 'I have a dream' and 'Yes, we can', these echoes where people accompany the main voice, like background singers.

A way with words has always been an asset, and, when used for political ends, or in times of crisis, it's worth a hundred battalions. In ancient Greece, Demosthenes employed his powerful oratorical skills trying to warn the citizens about the imminent danger of invasion by Philip of Macedon. When Philip advanced on Thrace, the Athenians called an assembly to debate whether or not to heed the great orator's advice. Demosthenes was sick of people taking liberty and the Athenian way of life for granted, and he boldly called upon the assembly to rise up and take action.

US Senator Barack Obama at the 'Yes, we can' rally in the Johnson Hall at George Mason University

It is this fate, I solemnly assure you, that I dread for you, when the time comes that you make your reckoning, and realize that there is no longer anything that can be done. May you never find yourselves, men of Athens, in such a position! Yet in any case, it were better to die ten thousand deaths, than to do anything out of servility towards Philip or to sacrifice any of those who speak for your good. A noble recompense did the people in Oreus receive, for entrusting themselves to Philip's friends, and thrusting Euphraeus aside! And a noble recompense the democracy of Eretria, for driving away your envoys, and surrendering to Cleitarchus! They are slaves, scourged and butchered.

After his rousing speech, the assembly all cried out, 'To arms! To arms!'

Brevity is also a great attribute. Perhaps the most famous speech in American history is Abraham Lincoln's Gettysburg Address. It was delivered on the site of the newly consecrated cemetery in Gettysburg, but ironically it was never intended to be the main event. The principal speech was by the former Secretary of State, Edward Everett, which lasted a good two hours and ran to nearly 14,000 words. Lincoln's, by contrast, lasted two minutes and was only 275 words long. Today, who remembers Everett's words?

The Gettysburg Address

Four score and seven years ago our fathers brought forth on this continent, a new nation, conceived in Liberty, and dedicated to the proposition that all men are created equal. Now we are engaged in a great civil war, testing whether that nation, or any nation so conceived and so dedicated, can long endure. We are met on a great battle-field of that war. We have come to dedicate a portion of that field, as a final resting place for those who here gave their lives that that nation might live. It is altogether fitting and proper that we should do this. But, in a larger sense, we

can not dedicate – we can not consecrate – we can not hallow
– this ground. The brave men, living and dead, who struggled
here, have consecrated it, far above our poor power to add or
detract. The world will little note, nor long remember what we
say here, but it can never forget what they did here. It is for us
the living, rather, to be dedicated here to the unfinished work
which they who fought here have thus far so nobly advanced.
It is rather for us to be here dedicated to the great task remaining
before us – that from these honored dead we take increased
devotion to that cause for which they gave the last full measure
of devotion – that we here highly resolve that these dead shall
not have died in vain – that this nation, under God, shall have
a new birth of freedom – and that government of the people,
by the people, for the people, shall not perish from the earth.

And then there are the speeches whose power lies in simplicity, in
quiet, heartfelt, searing honesty. In the 1870s, the Native American
leader Chief Joseph tried to resist his tribe of Nez Perce being moved
to a reservation by the US military. After months of fighting and a
forced march of 1,300 miles towards the Canadian border, he
surrendered with these words:

> I am tired of fighting. Our Chiefs are killed; Looking
> Glass is dead, Ta Hool Hool Shute is dead. The old men
> are all dead. It is the young men who say yes or no. He
> who led on the young men is dead. It is cold, and we have
> no blankets; the little children are freezing to death. My
> people, some of them, have run away to the hills, and have
> no blankets, no food. No one knows where they are – per-
> haps freezing to death. I want to have time to look for my
> children, and see how many of them I can find. Maybe I
> shall find them among the dead. Hear me, my Chiefs!
> I am tired; my heart is sick and sad. From where the sun
> now stands I will fight no more forever.

Chief Joseph's surrender speech may have immortalized him as a
great orator, but it did little to help his cause. The US military reneged

on their agreement, and Chief Joseph never returned to his homeland in Idaho.

One of the greatest of all orators was Winston Churchill. His use of the language, cadence, repetition and the honesty implied in the simplicity of the Anglo-Saxon words in his famous 'fight them on the beaches' speech can still raise goosebumps:

Chief Joseph became immortalized as a great orator

*We shall not flag or fail. We shall go on to the end, we shall
fight in France, we shall fight on the seas and oceans, we
shall fight with growing confidence and growing strength
in the air, we shall defend our Island, whatever the cost
may be, we shall fight on the beaches, we shall fight on the
landing grounds, we shall fight in the fields and in the
streets, we shall fight in the hills; we shall never surrender.*

The repetition of the same phrase – 'we shall fight' – is the same
device that Martin Luther King used in 'I have a dream.' And every
word from 'We shall defend' onwards (whether Churchill chose them
consciously or not) is Anglo-Saxon, apart from 'surrender', which
is French. Churchill's speech worked because it spoke to everyone.
As the American journalist Edward R. Murrow commented, 'He
mobilized the English language and sent it into battle.'

*Winston Churchill on the balcony of Buckingham Palace on
VE Day, 1945*

Propaganda

Propaganda, said the English academic Francis Cornford, is 'that branch of the art of lying which consists in very nearly deceiving your friends without quite deceiving your enemies'.

Advertising may have moved on since the Golden Days of Madison Avenue, but propaganda has never shied away from its primary role of explicit persuasion. In its most basic form, it's about disseminating information for any given cause, and it's been around for as long as we've had organized societies with messages to promote.

It doesn't mean it can't be entertaining or subtle. Some of the best British propaganda was produced in the dozens of films made during the Second World War. Of course, some of them are truly awful, cringe-making examples of pragmatic jingoism, but others have become classics, like the 1942 *Went the Day Well?* (based on a Graham Greene short story), which tells how a vanguard of Germans, disguised as British soldiers, infiltrates a sleepy, idyllic English village to prepare the way for invasion. The villagers eventually manage to fight back, with a combination of British pluck, ingenuity and surprising ruthlessness (in one scene the local postmistress kills a German with an axe), and Britain is saved. The purpose, of course, was to warn against the danger of invasion (which, in fact, by 1942 was not so real) and ram home the need for constant vigilance; to portray German brutality and evil and, above all, to extol British patriotism, sacrifice and bravery – all standard war propaganda.

But it's war propaganda done with some style and artistry, with touches of pathos, and the occasional satirical side-swipe at other enemies – like the great line, completely un-pc nowadays, of course, when a young village lad is being urged to do his part in defending the British Isles. 'You know what "morale" is, don't you?' someone asks him. 'Yes,' he declares, 'it's what the wops ain't got!' There's black humour, rousing British cheeriness and flashes of shocking realism which knock against the rural idyll cliché, all of which make it a more powerful piece of cinema, and therefore a more powerful piece of propaganda.

Propaganda can be blunt or subtle, blindingly obvious or relentlessly and cleverly suggestive, but its aim has always been to persuade and get everyone on message, whatever the method. It's been used by regimes to lie, dissemble, exhort, convert and cover up. Totalitarian regimes have created huge ministries of propaganda, while democracies have given birth to their bastard offspring – the PR companies and spin doctors.

Propaganda films were hugely popular during the Second World War

It's been wildly successful, but thankfully we usually end up seeing through it. In works of fiction like George Orwell's *1984*, the propaganda ministry realizes that to really change people's minds it needs to change the language itself. If you can't say it, you won't think or feel it. So the Newspeak dictionary gets slimmer and slimmer as more and more words are 'disappeared'. Newspeak, where only prescribed words can be used, is the propagandists' dream.

Newspeak

'It's a beautiful thing, the destruction of words' (1984)

Newspeak (pronounced New-speak) is the fictional language created by George Orwell in his novel *1984*. It's closely based on English but with a vastly reduced and simplified vocabulary and grammar. The aim of the totalitarian government of Oceania is to make language so basic, so sterile that any alternative way of thinking to that prescribed by the regime (thoughtcrime) is impossible. In a nightmarish twist on Sapir–Whorfism, how can you have ideas of freedom or rebellion if there are no words for them?

'In the end we shall make thoughtcrime literally impossible,' explains the character Syme, a Newspeak specialist, 'because there will be no words in which to express it ... [We're] cutting the language down to the bone . . . Newspeak is the only language in the world whose vocabulary gets smaller every year.'

Richness of language is anathema to the Party. Words must be stripped of superlatives or negatives and reduced to the absolute minimum. 'Uncold' for warm; 'unlight' for dark. Good and evil and all nuances in between are distilled into six words: good, plus good, doubleplusgood, ungood, plusungood and doubleplusungood.

Many of Orwell's Newspeak words appear to be modelled on Esperanto. For instance, in Esperanto the opposite of good, *bona*, is *malbona*, and its extreme is *malbonega* – literally 'very ungood'. When he travelled to Paris in 1928 for his 'down and out' period, Orwell is thought to have stayed with his aunt Nellie and her partner, Eugène Lanti, a leading Esperantist. According to his biographer Bernard Crick, Orwell told a friend that he had gone to Paris 'partly to improve his French', but had to leave his first lodgings because the landlord and his wife only spoke Esperanto – 'and it was an ideology, not just a language'. So, whether out of antipathy to Esperanto or a fascination with a language constructed for political aims, Orwell was clearly inspired.

'Thoughtcrime' would be impossible and everyone would speak 'Newspeak' in Orwell's 1984

In *1984*, Newspeak was still being introduced as an official language, but the expectation was that Oldspeak, the Newspeak term for English, would have disappeared by 2050; all literature from the past – apart from technical manuals – would be unintelligible and untranslatable. The 'all men are equal' passage from the Declaration of Independence would be reduced to a single word: crimethink.

Orwell's *1984* introduced an astonishing number of words and concepts denoting totalitarian authority into the English language – Big Brother, the Thought Police, Room 101, doublethink, unperson, thoughtcrime. Chilling slogans resonating as clearly today as they did over sixty years ago – 'Who controls the past controls the future; who controls the present controls the past.'

It's an idea far more powerful and frightening than the most destructive of weapons: loss of language leads to loss of our history and, ultimately, loss of our selves.

In the Western liberal tradition we're pretty sniffy about overt propaganda – for us, it's a negative, dishonest tool – but other systems and ideologies see it as a positive means of motivating the population and keeping it under control.

One of the first modern propagandists was Josef Goebbels, who became Hitler's Reich Minister for *Volksaufklärung und Propaganda* (Public Enlightenment and Propaganda). It's chillingly ironic that he should use the inference of the Enlightenment in his title when he was the master of the dark arts, but then 'Arbeit Macht Frei' ('Work Makes You Free'), the slogan that adorned the entrance to the extermination camps, was hardly a promise they kept.

As editor of *Der Angriff* (*The Attack*) Goebbels masterminded Hitler's attacks on his opponents, which were particularly effective in the election of 1932, when Hitler become Chancellor. He used radio, cinema and spectacular demonstrations, which culminated in the famous Nuremberg rally. His showmanship helped to get the Nazi message across and then keep the German people firmly corralled within the Nazi ideology, but his techniques were just as often crude. He was interested in effectiveness, not art. 'Good propaganda is successful,' he declared. 'Bad fails to achieve its desired result.'

To this end Goebbels exploited racism, xenophobia and envy to the limit and sold the idea of the master race and the thousand-year Reich. His power and influence dwindled the more successful the Nazi Party was in extinguishing any opposition. There was simply less need for that sort of propaganda when no one dared complain or criticize, and by the beginning of the war his main task had been accomplished. What he did understand, however, was the need for the German people to have some light relief from the rigours of war and he became the David Selznick of the Nazi movie industry, producing an endless supply of frothy comedies and romances from the famous UFA studios on the outskirts of Berlin. It's hard to imagine Nazi comedies now.

The propaganda – the control of what people were to believe, and the way they were to live – went far beyond overt messages and the production of films. Goebbels' influence was pervasive.

Holiday complexes and cruise ships were built for the Nazi worker. The KdF (*Kraft durch Freude* – Strength through Joy) became the largest tourist operator in the world, and Goebbels' ministry was even given the task of enticing foreign tourists to spend their holidays in Germany. By 1939 over 25 million people had been on KdF holidays – though, unsurprisingly, few came from overseas. KdF also built the first VW Beetle and marketed it as the workers' car – a subtle blend of propaganda and advertising.

The thing about propaganda is, it doesn't matter what the message and purpose is – it could be anything – the point is finding ways of promoting it and ensuring that people believe it. Take Stalin in the Soviet Union and Chairman Mao in China, masters of propaganda both – they could give Goebbels a run for his money. They used similar methods to the Nazis, but they were promising different results: a utopian socialist society, rather than the thousand-year Reich.

Adolf Hitler demonstrated impressive showmanship

All these dictatorships made extensive use of the printed word, speeches and, especially, posters. Stalin's ambitious Five Year Plans, with their slogans exhorting workers to up their quotas and defeat the imperialists, were none too subtle, but the images of a smiling benevolent Uncle Joe did much to win the proletarian heart. In China, propaganda was taken to a new level with Mao's continual revolutions within the revolution, each new wave designed to direct and control, under the guise of improving the nation and the lives of the people.

The Hundred Flowers Movement of 1957, for example, was a short-lived experiment in asking intellectuals to criticize the direction of the revolution. 'Letting a hundred flowers blossom and a hundred schools of thought contend is the policy for promoting progress in the arts and the sciences and a flourishing socialist culture in our land,' was Mao's initial premise. But when the criticisms became personal attacks on him he ruthlessly squashed it and effectively killed off any further intellectual debate.

German propaganda was explicit and forceful. This example compares healthy German children with hungry, barefoot Russian children

Propaganda turns language upside down. It twists meaning and says the opposite of what it does. (It's all in *1984* – have a look again. 'War is peace. Freedom is slavery. Ignorance is strength.') In his attempts to neutralize any further division within the party, Mao launched 'The Great Leap Forward' in 1958. Propaganda was used extensively to mobilize the masses and change the country from a predominantly agrarian economy to an industrial one. But there was no leap forward. The opposite happened, with catastrophic results. An estimated 20–40 million people died of starvation as the economy went into freefall.

And that's where another side of propaganda kicks in. When the reality is not as the propaganda promises, just keep pumping out the message until people are swamped. Mao's 'Little Red Book' was one of the tools of revolution which flooded the consciousness of millions of citizens with the wisdom and pithy sayings of the Chairman. It was impossible to escape the message. The People's Liberation Army General Political Department printed a total of 1,005,549,800 copies of the book, in sixty-five languages. That's over a billion 'hits' of propaganda in a pre-internet age.

Chairman Mao with peasants during the Cultural Revolution of 1966

Of course, as well as making some fine war films, Western society has put out its fair share of outright wartime propaganda. Lord Kitchener's recruitment campaign during the First World War was more direct than anything dreamed up by the communist propagandists, and Uncle Sam was no slouch either. Today's messages, though, are couched in subtler and more entertaining ways. The US Defense Department, through the US Army's Office of Economic and Manpower Analysis, has developed its own video game, 'America's Army', to try to lure new recruits. Available as a free download, it's been a huge success with over 8 million registrations.

Today's politicians in general, though, have a much tougher job getting their message across in what the PR chaps would call 'a crowded marketplace'. As ad man Don Bowen says, 'Political slogans are still useful because they sum up the approach that the political party is moving in. The difficulty is there's always somebody with an antithetical political statement that they're trying to get over. So, unlike propaganda, when there's only one story in town, which was true in Nazi Germany and in Mao's China, here there are many stories. We have choices. There isn't just one beer that says, "Probably the best lager in the world," there's another beer that says, "Refreshes the parts other beers cannot reach." And there's another beer that's saying something else. So you never just get your own way.'

Well, that's democracy for you. And in a democracy, people are free to pick and choose, and there are many, many different voices we can listen to, or ignore. And, generally, we prefer the simpler message. Don't make it too complicated, or it'll just get lost. Hitler knew that very well. 'The most brilliant propagandist technique will yield no success unless one fundamental principle is borne in mind constantly,' he said. 'It must confine itself to a few points and repeat them over and over.' Which is as true today as then. A slogan like Prime Minister David Cameron's 'Big Society' struggles to be heard, so little wonder he has to repeat it every time he's on the radio or TV.

Does ultimate resistance to the power of state propaganda depend on there being so many channels of communication, so many messages, that no one authority or idea can dominate? It's true that even within totalitarian regimes the profusion of social network sites

'Stalin Leads Us to Victory', 1943

like Twitter and Facebook is chipping away at their hegemony. The events in the Middle East and the Maghreb in 2011 show just how difficult it is to contain information and channel people's thoughts. When a young street seller in Tunisia sets himself alight because he can no longer fight the system which makes scraping a living almost impossible, and a very short time later his story is texted and tweeted and posted round the world – that's when ideas can't be corralled. When so-called 'citizen journalists' can upload pictures and videos of events to the internet in a matter of minutes, then the windows of communication are well and truly open.

We can now share our ideas, our anger, our frustrations with thousands of people at the touch of a button. When the financial crash in Iceland happened in 2008, it was a Facebook campaign by ordinary Icelanders that ultimately persuaded the government to support an investigation into the people responsible for the banking collapse. If the government wouldn't take action, the people would. Goebbels would not have liked that. And could he possibly have imagined a world where WikiLeaks existed, a world where he couldn't control the message, because, with a click of a mouse, someone, somewhere with access to government secrets and confidential documents can upload them to the web?

The language of Twitter, texting and social network sites may not be the most elegant, but it shows that we humans have this innate need to communicate, be it banal gossip, love stories or life-changing revolution.

The Future

So how will the storytellers of the digital age gain and keep our attention? Will language still be as powerful as it is today? Will it be controlled by an elite of media owners, the neo-Brahmins of the twenty-first century, or democratized so that we all become producers rather than consumers? What does the future look like on Planet Word?

In an elegant, light-filled building worthy of a sci-fi movie, MIT's elegant new Media Lab houses a division called 'The Future of Story-telling'. The idea is tantalizingly simple and almost impossible to envisage. Yes, we will undoubtedly have immersive 3D screens, and video games that involve physically wearing a suit to mimic the movements in the game (because they're already invented). And, yes, we'll be able to roll up a screen like a newspaper and put it in our pocket, and have adverts directed at us personally as we walk around the cities of the future (like in Spielberg's movie *Minority Report*). And, yes, we'll be able to create stories across space with interactive screens linked to the web: families will be able to create their own histories or mythologies, to preserve images and voices for future generations. But will all this new technology change what we want to say?

Communications theorist Marshall Mcluhan was wrong: the medium is really not the message. The message is the message. Humans will continue to be addicted to chatter and gossip, to talk of love and sex and sometimes of death. We will go on flirting and cracking jokes and making up wild stories that have no sense. In other words, the same things we talked about when we all sat around a fire at the end of the day, roasting a zebra or slab of sperm whale. What is different is that the global village which Mcluhan predicted in the early 1960s is now a reality, and the realization of it has been the biggest transformation in our species' evolution since . . . well, maybe the invention of language itself. What is certain is there will be many more stories to tell.

Notes

Chapter 1

'confounded their language' *Genesis*, 11:7

'the greatest wonder in England' *The Literary Gazette and Journal of the Belles Lettres, arts, sciences &c,* James Moyes, 1829

'answering questions, telling the hour of the day' Charles Dickens, as quoted in *Jay's Journal of Anomolies, Conjurers, Cheats, Hustlers, Hoaxsters, Pranksters, Jokesters, Imposters, Pretenders, Side-Show Showmen, Armless Calligraphers, Mechanical Marvels, Popular Entertainments,* Farrar, Straus and Giroux, 2001

'The greatest curiosity of the present day' Billboard for Toby the Sapient Pig, 1817

'. . . spell and read, play at cards' Billboard for Toby the Sapient Pig, 1817

'A program of research' Dwight 'Wayne' Batteau, *Man/Dolphin Communication Final Report* 1966–1967

'The dolphins were able to' Louis Herman, *Cognition*, Herman, Richards, & Wolz, 1984

'Their capacity for communication' Robert Frederking as quoted in 'Whatever happened to . . . talking dolphins', Susan Kruglinski, *Discover* magazine, 2006

'. . . about as likely that an ape' Noam Chomsky, interviewed by Matt Aames Cucchiaro in 'On the Myth of Ape Language', email correspondence 2007–8

'How do you reconcile a tiny chimp' Jenny Lee, as quoted in *Nim Chimpsky: The Chimp Who Would Be Human*, Elizabeth Hess, Random House, 2008

'Put the pine needles in the refrigerator' Sue Savage-Rumbaugh, *Kanzi and novel sentences* video, greatapetrust.org

'I used to think my aim' Dr Cathy Price, *Fry's Planet Word*, BBC 2011

'Language Acquisition Device' (LAD), Noam Chomsky, *Aspects of the Theory of Syntax*, MIT Press, 1965

'Human communication' Michael Tomasello, *Origins of Human Communication*, MIT Press, Cambridge, MA, 2008

'I'm at the front of the lecture hall' ibid.

'This is hopefully the first' Dr Wolfgang Enard, *BBC News*, 2002

'Language at a bare minimum' Steven Pinker, *Fry's Planet Word*, BBC 2011

'. . . there also has to be some kind of talent' ibid.

'"All gone sticky" . . . Now that doesn't correspond' Steven Pinker, *Fry's Planet Word*, BBC 2011

'"More outside" . . . .That's quite a cognitive feat' ibid.

'"He sticked it on the paper", 'He teared the paper", "We holded the baby rabbits" ibid.

'It's an extremely powerful' ibid.

'Jean: Okay, Okay, now this is another creature, this one's called a tass. That's a tass' Jean Berko Gleason, *Fry's Planet Word*, BBC 2011

'Young kids' brains are not formed' ibid.

'foster mothers and nurses' Salimbene di Adam, *Chronicle of Salimbene De Adam* edited by J.L Baird, G Baglivi and J.R. Kane *(Medieval and Renaissance Texts and Studies)*, Binghamton, NY, April 1986

'Some say they spoke good Hebrew' Robert Lyndsay of Pitscottie as quoted in *Old and New Edinburgh*, James Grant, Cassell & Co, 1880

'It is more likely they would scream' Sir Walter Scott, as quoted in 'The Bird Man of Stirling', *BBC History*

'First he embraced her with his armes' Vicar of St Martin's Church, *Tudor Era*, Headline History website

'[The] two, when they chanced to meet' Richard Carew, *Survey of Cornwall 1602*, Mark Press, 2000

'there comes in that Dumb boy' Samuel Pepys, *The Diaries of Samuel Pepys*, Penguin Classics, 2003

'The deaf tend' Janiece Brotton, *Fry's Planet Word*, BBC 2011

'Syntax, the constraints on language' Judy Shepherd-Kegl, *Fry's Planet Word*, BBC 2011

'Can't express your feelings' Adrian Perez, CBS News, 2009

'The single gesture doesn't have rhythm' Judy Shepherd-Kegyl, *Fry's Planet Word*, BBC 2011

'If you're beyond the critical period' ibid.

'You know, we can look back' ibid.

'If you have children you've had the experience' ibid.

'To have a second language is to have a second soul' Charlemagne, Holy Roman Emperor

'I speak French to my ambassadors' Frederick the Great of Prussia

'(language) is not merely a reproducing instrument' Benjamin Whorf, *Language, Thought and Reality: Selected Writings of Benjamin Lee Whorf*, MIT Press, 1964

'When my wishes conflict with my family's' As quoted by Lee Gardenswartz and Anita Rowe, *Managing Diversity: A Complete Desk Reference and Planning Guide*, Jossey Bass, 1993

'In English, we tend to divide our space' Lera Boroditsky, *Fry's Planet Word*, BBC 2011.

'Chicken and egg isn't the right way' ibid.

'Nearly all of my labours' Jacob Grimm, *Selbstbiographie*, from *Kleinere Schriften Vol 1* F. Dümmler, Berlin, 1864

'To réecs éhest ("Once there was a king")' J. P. Mallory and D. G. Adams, *Encyclopaedia of Indo-European Culture*, Fitzroy Dearbon Publishers, 1997

'Oh no. No, no, no, no!' Zaha Bustema, *Fry's Planet Word*, BBC 2011

Chapter 2

'Learn well the language of the whites' Anonymous, Hawaii, 1896

'In the old days,' Aju, *Fry's Planet Word*, BBC 2011

'At my home I speak Akha' Aju, *Fry's Planet Word*, BBC 2011

'Irish golfer', *Fry's Planet Word*, BBC 2011

'We next reached Skibbereen . . .' James Mahony, *The Illustrated London News*, 1847

'I have been assured' Jonathan Swift, *Modest Proposal (for Preventing the Children of Poor People in Ireland Being a Burden to Their Parents or Country, and for Making Them Beneficial to the Public)*, Prometheus Books UK, September 1995

'If we allow one of the finest' Douglas Hyde, *An leabhar sgeulaigheachta*, Dublin Gill, 1889

'I would earnestly appeal' Douglas Hyde, *The Necessity for De-anglicising Ireland*, Academic Press, Leiden, 1994

'The Gaelic League restored the language to its place' Michael Collins, *The Path To Freedom*, Talbot Press Ltd, Dublin 1922

'to build up a Celtic and Irish school of dramatic literature' Lady Augusta Persse Gregory, *Our Irish Theatre: A Chapter of Autobiography*, GP Putnam's Sons, The Knickerbocker Press, New York and London, 1913

'as calm and collected as Queen Victoria' As quoted by Carmel Joyce, *Inspirational Figures from Irish History (Lady Isabella Augusta Gregory, Samuel Beckett, playwrights)*, World of Hibernia, 2000

'You have to remember that Irish as a language was spoken up till the 1840s and 50s by many, perhaps the majority of Irish people' Declan Kiberd, *Fry's Planet Word*, BBC 2011

'I know that if you were to speak' Hugh Farley, *Fry's Planet Word*, BBC 2011

'I think that we have as a nation' ibid.

'"*nam alii oc, alii si, alii vero dicunt oil*"' ('Some say *oc*, others say *si*, others say *oil*"') Dante, *De Vulgari Eloquentia* (Cambridge Medieval Classics), translated by Steve Botterill, Cambridge University Press, 1996

'The monarchy had reasons to resemble the Tower of Babel' Bertrand Barère, National Convention, 1794

'My grandparents speak Breton too' Nicolas de la Casiniere, *Ecoles Diwan, La Bosse du Breton,* 1998

'Instead of setting their ambitions' Frédéric Mistral, *Speech to the Félibres of Catalonia* as quoted by Jennifer Michael *in Journal of American Folklore 111,* American Folklore Society, 1998

'in recognition of the fresh originality' Frédéric Mistral, The Nobel Foundation, 1904

'I was profoundly shocked' President Chirac, EU Summit, March 2006, as reported in Nicholas Watt and David Gow, 'Chirac vows to fight growing use of English', *Guardian,* 25 March 2006

'. . . because that is the accepted business language of Europe today' Ernest-Antoine Seilliere, as quoted in 'Chirac upset by English address', BBC News, 24 March 2006

'But you know, what they, the regional languages, have lost is not too much' Marc Fumaroli, *Fry's Planet Word,* BBC 2011

'Jesus said, "Love is everything"' Dr Ghil'ad Zuckermann, *Fry's Planet Word,* BBC 2011

'Israeli is a very complex language' ibid.

'The Hebrew language can live only if we revive the nation' Eliezer Ben Yehuda as quoted in *Jew! Speak Hebrew,* Aliyon, 2005, on the Jewish Agency for Israel website www.jafi.org.il

'In a heavy atmosphere' David Yudeleviz, as quoted by Ilan Stavans 'Crusoe in Israel', *Pakn Treger: The magazine of the Yiddish Book Center,* No. 58, December 2009, yiddishbookcenter.org

'Yiddish speaks itself beneath Israeli' Dr Ghil'ad Zuckermann, *Fry's Planet Word,* BBC 2011

'Yiddish has not yet said its last word' Isaac Bashevis Singer, *Nobel Lectures 1978,* Literature 1968–80, World Scientific Publishing Company, Singapore, 1993

'There is one people – Jews . . . and its language is – Yiddish' I. L. Peretz, quoted in *Dr Birnboym's Vokhnblat #2,* Czernowitz, 1908 (translated by Marvin Zuckerman and Marion Herbst from *Nakhmen Meisel, Briv unredes fun,* YKUF, New York, 1944)

'It's a great language for saying the man's a dick' Stewie Stone, *Fry's Planet Word,* BBC 2011

'The *nebach* is the guy getting the water when you're going on a football team' Ari Teman, *Fry's Planet Word,* BBC 2011

'For an older Jewish audience' Stewie Stone, *Fry's Planet Word,* BBC 2011

'I sent a rabbi a joke' Ari Teman, *Fry's Planet Word,* BBC 2011

'When one Jew meets another Jew' Stewie Stone, *Fry's Planet Word,* BBC 2011

'Yiddish is basically our soul' ibid.

'I grew up, and I was *bah mitzvahed*' ibid.

'Hebrew comes from the vocal chords' ibid.

'a serous and watery purgative motion' John Wilkins, *An Essay Towards a Real Character and a Philosophical Language,* The Royal Society, 1668, Thoemmes Continuum, Facsimile Ed edition, January 2002

'I was taught that all men were brothers' Ludovic Lazarus Zamenhof in a letter to N. Borovko, 1895, quoted by David Poulson in *Origin of Esperanto* July 1998

'We believe it is a language' Littlewoods, BBC News, July 2008

'If you lose a contract to a Moroccan rival' Jean Paul Nerriere, *Parlez Globish,* Eyrolles, 2006

'It's a proletarian and popular idiom' ibid.

'The only jokes which cross frontiers' Jean Paul Nerriere, as quoted by Adam Sage, 'Globish cuts English down to size', *The Times*, December 2006

'I am helping the rescue of French' Jean Paul Nerriere, *Parlez Globish,* Eyrolles, April 2006

'To make up a non-human language' Mark Okrand, *Fry's Planet Word,* BBC 2011

'because there's lots of zees and zots in science fiction' ibid.

'We arrived on the set one day' ibid.

'You have not experienced Shakespeare' ibid.

'My son was born in 1994' ibid.

'Well, he was learning it' Dr d'Ormond Speers, *Fry's Planet Word,* BBC 2011

'I would say something to him in Klingon' ibid.

'The reason this language is so successful' Arika Okrent, *Fry's Planet Word*, BBC 2011

'Well, it's given me a deeper appreciation' ibid.

'A language is a dialect with an army and a navy' Max Weinreich, *Der yivo un di problemen fun undzer tsayt,* New York, 1945

'long *es* and very slow'; 'they don't say their *ts*' Ian McMillan, *Fry's Planet Word,* BBC 2011

'harsh . . . it's to do with the harsh winds' ibid.

'That's where an isogloss happens' ibid.

'My Aunty Mabel, who was from Chesterfield' ibid.

'Whenever I speak in my voice' ibid.

'So, somehow, the language carries on' ibid.

'When the word dies' Ian McMillan, 'Utopia! If you frame thissen properly, that is', *Yorkshire Post*, 23 August 2010

'And when I was first on the radio' Ian McMillan, *Fry's Planet Word*, BBC 2011

'It is the business of educated people' Arthur Burrell, as quoted in the *Journal of International Phonetic Association*, 1987, p. 21

'One hears the most appalling travesties' John Reith, *Broadcast Over Britain*, Hodder and Stoughton, London, 1924

'be able to recognize instantly' BBC 1940

'And to all in the North, good neet' Wilfred Pickles, BBC Radio, c.1940

'It is impossible for an Englishman' George Bernard Shaw, *Pygmalion (Preface to Pygmalion)* Penguin Classics, revised edition, January 2003

'Having one's cards engraved . . .' Professor Alan Ross in Nancy Mitford (ed.), *Noblesse Oblige: An Enquiry into the Identifiable Characteristics of the English Aristocracy*, 1956

'Phone for the fish knives, Norman' John Betjeman in Nancy Mitford (ed.), *Noblesse Oblige: An Enquiry into the Identifiable Characteristics of the English Aristocracy*, Hamish Hamilton, 1956

'Can a non-U-speaker become a U-speaker?' Professor Alan Ross in Nancy Mitford (ed.), *Noblesse Oblige: An Enquiry into the Identifiable Characteristics of the English Aristocracy*, Hamish Hamilton, 1956

'Why should they hide it?' Lawrence Fenley, *Fry's Planet Word*, BBC 2011

'In the twenty-first century' ibid.

'It is only when, usually, you have an issue' ibid.

'May it be forbidden that we should ever speak like BBC announcers' Wilfred Pickles, *Between You and Me*, Werner Laurie. Reprinted by permission of The Random House Group Ltd

'Barnsley's what I think with' Ian McMillan, *Fry's Planet Word*, BBC 2011

Chapter 3

'Those are the heavy seven' George Carlin, 'Seven Words You Can Never Say on Television', *Class Clown*, Atlantic, 1972

'But words are words. I never did hear / That the bruised heart was pierced through the ear' William Shakespeare, *Othello*, (1:2, 218–19) Penguin Classics, New Ed edition, 2005

'Not one of them would sit down,' Captain James Cook, *James Cook: The Journals*, Penguin Classics, 2003

'I would hear things' Timothy Jay, BBC, *Fry's Planet Word*, 2011

'It's like using the horn on your car' ibid.

'As soon as kids can speak, they're using swear words' Professor Timothy Jay, *Cursing in America*, John Benjamins, Philadelphia, 1992

'Cathartic swearing comes from a primal rage circuit' Steven Pinker, *The Stuff of Thought*, Allen Lane, 2007

'I had noises' Jess Thom, *Fry's Planet Word*, BBC 2011

'It's going all the time biscuit' ibid.

'I was speaking to my dad on the phone' ibid.

'Absolutely. I think lots of people misunderstand Tourette's' ibid.

'The response is not only emotional but involuntary' Steven Pinker, *Why We Curse: What the F***? New Republic* magazine, August 2007

'One of mankind's greatest-ever living language centres' Peter Silverton, *Filthy English: The How, Why, When and What of Everyday Swearing*, Portobello Books, 2009

'In my mind I'd say that' Les Duhigg, *Fry's Planet Word*, BBC 2011

'I was always apologizing for him' Marion Duhigg, *Fry's Planet Word*, BBC 2011

'We had a chap in the stroke group' ibid.

'It's like being born again' Les Duhigg, *Fry's Planet Word*, BBC 2011

'*Richard*: So, Stephen, when you put your hand in the water . . .

'*Stephen*: That is cold actually . . . ' Dr Richard Stephens, Brian Blessed and Stephen Fry, *Fry's Planet Word*, BBC 2011

'Please could you take this note, ram it up his hairy inbox and pin it to his fucking prostrate' Armando Ianucci, *The Thick of It: The Rise of the Nutters,* BBC 2007

'There was that world which lived off a twenty-four-hour news cycle' Armando Ianucci, *Fry's Planet Word*, BBC 2011

'The last thing I want is every programme' ibid.

'Euphemism is such a pervasive human phenomenon' Joseph Williams, quoted in Ralph Keyes, *Euphemania: Our Love Affair With Euphemisms*, Little Brown, 2010

'This Earl of Oxford making his low obeisance' John Aubrey, *Brief Lives,* Penguin, new edition, 1972

'Not a man swears but pays his twelve pence' Oliver Cromwell, quoted in *An Anatomy of Swearing*, Ashley Montague, University of Pensylvannia Press, March 2001

'I'd like breast' Winston Churchill, Virginia, 1929, quoted by Celia Sandys, *Chasing Churchill: Travels With Winston Churchill*, Harper Collins, new edition, 2004

'We had an auxiliary who was Portuguese' Julia Saunders, *Fry's Planet Word*, BBC 2011

'Ah, isn't that nice' Harry Carpenter at the 1977 Oxford–Cambridge boat race

'It hangs like flax on a distaff' William Shakespeare, *Twelfth Night*, 1:3,16, Penguin Classics, new edition, 2005

'They don't pay their sixpences' Marie Lloyd, quoted in *The New York Telegraph*, 14 November 1897

'... significant moment in English history' T. S. Eliot, *Selected Essays*, Faber and Faber, London, 1941

'When roses are red' Max Miller quoted in John M. East, *Max Miller: The Cheeky Chappie*, Robson Books Ltd, new edition, 1998

'I know exactly what you are saying' ibid.

'Programmes must at all cost be kept free of crudities' *The Little Green Book*, BBC 1949

'In Hackney Wick there lives a lass' Barry Took and Marty Feldman, 'Rambling Syd Rumpo', *Round The Horne*, BBC Radio

'Mrs Slocombe: Before we go any further, Mr Rumbold' David Croft and Jeremy Lloyd, *Are You Being Served?*: 'Our Figures Are Slipping', series 1, episode 3, BBC 1973

'The twittering of the birds all day, the bumblebees at play' Ronnie Barker, *The Two Ronnies*, BBC

'I spend all day just crawling through the grass' Peter Brewis, 'The Two Ninnies', *Not The Nine O'Clock News*, 1982

'When, 20 years ago, Molly Sugden' David Baddiel

She was pleased to see his tender won' *I'm Sorry, I Haven't A Clue*, BBC Radio 4

'I know, for example, that the lovely Farad here' Omid Djalili, *Fry's Planet Word*, BBC 2011

'My parents often had English people around' ibid.

'I just said thank you very much to Farad' ibid.

'I am disconsolate and nothing can cure me of my misery' E. M. Forster, *Arbinger Harvest: Notes on the English Character*, Mariner Books, January 1950

'My wife, who's British, said' Omid Djalili, *Fry's Planet Word*, BBC 2011

'To be called a bald, fat fart to your face' ibid.

'You couldn't operate on a yacht' Matt Allen, *Fry's Planet Word*, BBC 2011

'It strikes everyone as an extreme case' A. P. Rossiter, *Our Living Language*, Longman's Green & Co., 1953

'But do not give it to a lawyer's clerk to write' Miguel de Cervantes, *Don Quixote*, Vintage Classics, new edition, 2007

'Upon any such default' Security Agreement, Harbour Equity Partners, LLC, November 2010

'The physical progressing of building cases' Sir Ernest Gowers, *The Complete Plain Words*, Her Majesty's Stationery Office, 1955

'It may be said that no harm is done' ibid.

'good and useful ...' Sir Ernest Gowers, *The Complete Plain Words*, Her Majesty's Stationery Office, 1955

'is the verbal sleight of hand' David Lehman, *Sign of the Times: Destruction and Fall of Paul De Man*, Andre Deutsch Ltd, 1991

'Proactive, self-starting facilitator required' quoted by Christopher Howse, 'At the end of the day, you've given 110 per cent', *Telegraph*, 14 June 2007

'Using language as a way of obscuring the truth' Ian Hislop, *Fry's Planet Word*, BBC 2011

'It starts in management consultancies' ibid.

'What amuses me is the same management' ibid.

'Doublespeak is a language which pretends to communicate' William Lutz, *The New Doublespeak*, Harper Collins, 1996

'unlawful and arbitrary deprivation of life' *US State Department annual report*, 1984

'Nazism permeated the flesh and blood of the people' Victor Klamperer, *Lingua Tertii Imperii (The Language of the Third Reich)*, Continuum International Publishing Group, 2006

'the barbed wire was not facing the West' Gunter Böhnke, *Fry's Planet Word*, BBC 2011

'My mother lost her purse' ibid.

'Every joke is a tiny revolution' George Orwell, *'Funny But Not Vulgar' and Other Selected Essays and Journalism*, The Folio Society, 1998

'Erich Honecker arrives at his office early one morning' Florian Henckel von Donnersmarck, *The Lives of Others*, Buena Vista Pictures, 2006

'We're in a no-win, damned if you do and damned if you don't scenario' Peter Jackson, quoted in 'Jackson Talks Dam Busters: Controversial decision looms for WWII remake', IGN website, 6 September 2006

'David Howard should not have quit' Julian Bond quoted in Donald Demarco, 'Acting Niggardly', *Social Justice Review* 91, No. 3–4, March–April 2000

'I'd go to school on the Monday' Stephen K. Amos, *Fry's Planet Word*, BBC 2011

'When people say political correctness has gone mad' ibid.

'language of a highly colloquial type' *The Concise Oxford Dictionary of Current English*, adapted by H. W. Fowler and F. G. Fowler, First Edition, 1911

'special vocabulary of tramps or thieves' 1756, *Online Etymology Dictionary*, etymonline.com

'the dirtiest dregs of the wandering beggars' Alexander Gil as quoted by Henry Hitchens in *The Language Wars: A History of Proper English*, John Murray, February 2011

'that poisonous and most stinking ulcer of our state' Alexander Gil, *Logonomia Anglica*, Scolar Press, reissue of 1621 edition, February 1969

'the continual corruption of our English tongue' Jonathan Swift, *Tatler*,
No. 230, September 1710
'the choice of certain words' ibid.
'an epithet which in the English vulgar language' Francis Grose, *A*
Classical Dictionary of the Vulgar Tongue, 1785, University of
Michigan Library, April 2009
'The freedom of thought and speech' ibid.
'I have never seen a man of more original observation' Robert Burns, in a
letter to Mrs Dunlop from Ellisland on 17 July 1789, as quoted by
Jennifer Orr, BBC website, *Robert Burns*
'low habits, general improvidence . . .' John Camden Hotton, *A*
Dictionary of Slang, Cant and Vulgar Words, Taylor and Greening,
1860
'I likes a top o' reeb' Henry Mayhew, 'London Labour and the London
Poor', *The Morning Chronicle*, 1851
'the wandering tribes of London'; 'There exists in London a singular tribe
of men' John Camden Hotton, *A Dictionary Of Modern Slang, Cant,*
And Vulgar Words, Taylor and Greening, 1860
'harristocrats of the streets' ibid.
'A citizen of London, being in the country' Francis Grose , *A Classical*
Dictionary of the Vulgar Tongue, 1785, University of Michigan Library,
April 2009
'one born within the sound of Bow bell, that is in the City of London'
John Minsheu, *Ductor in Linguas (Guide into the Tongues) and*
Vocabularium Hispanicolatinum (A Most Copious Spanish Dictionary)
(1617), Scholars Facsimilies & Reprint, May 1999
'Yankee Doodle came to London, just to ride the ponies' George M.
Cohen, 'Yankee Doodle Dandy', 1942
'pulling someone's pants up sharply to wedge them between the buttocks'
Jonathan E. Lighter, *Historical Dictionary of American Slang, Vol. 2:*
H–O, Random House Reference, 1997
'Oh, it's the tourists . . . I'm not Listerine but they get on my goat' Stephen
Fry, *Fry's Planet Word*, BBC 2011
'Well, we're losing it, aren't we?' London Cab Driver, *Fry's Planet Word*,
BBC 2011
'A Cockney has got a cheerful way about him' ibid.
'As feely ommes, we would zhoosh our riah' Peter Burton, *Parallel Lives*,
Gay Men's Press, 1985
'Hello. Is there anybody there?' Barry Took and Marty Feldman, 'Julian
and Sandy', *Round the Horne*, BBC Radio
'Omes and palones of the jury' Barry Took and Marty Feldman, 'Bona
Law', *Round the Horne*, BBC Radio

'a miracle of dexterity at the cottage upright' Barry Took and Marty
Feldman, 'Julian and Sandy: Bona Performers', *Round the Horne*, BBC
Radio

'In the beginning, Gloria created the heaven and the earth' The Sisters of
Perpetual Indulgence, King James Bible into Polari in 2003, as quoted
by Christopher Bryant, *Paul Baker: How Bona to Vada Your Dolly Old
Eek, Polari Magazine*, 3 December 2008

'A lot of English people see Australians as a recessive gene' Kathy Lette,
Fry's Planet Word, BBC 2011

'This is accounted for by the number of individuals' Peter Cunningham,
Two Years in New South Wales, Henry Colburn, 1827

'Pommy is supposed to be short for pomegranate' D. H. Lawrence,
Kangaroo, Penguin Classics, new edition, 1986

'I think it's something to do with our Irish heritage' Kathy Lette, *Fry's
Planet Word*, BBC 2011

'We shorten everything' ibid.

'I've met six British prime ministers' Stephen Fry, *Fry's Planet Word*,
BBC 2011

'having a black belt in tongue-fu' Kathy Lette, *Fry's Planet Word*,
BBC 2011

'gutless spivs' Australian Prime Minister Paul Keating, quoted by Nick
Stace in 'Taking Insults to a New Level', *My Telegraph*, September 2010

'brain-damaged'; 'mangy maggot'; 'the little dessicated coconut'
Australian Prime Minister Paul Keating, quoted by Patrick Carlyon in
'Ex-PM Paul Keating the heckler we had to have', *Herald Sun*,
3 November 2009

'like being savaged by a dead sheep' Denis Healey on Sir Geoffrey Howe,
Hansard, 14 June 1978, col. 1027

'into the mysteries of Australian colloquial speech' Barry Humphries 'The
Adventures of Barry McKenzie', *Private Eye*, c.1960

'Barry is hugely observant' John Clarke, *Fry's Planet Word*, BBC 2011

'relied on indecency for its humour' Australian Department of Customs
and Excise, quoted in *The Mythical Australian: Barry Humphries,
Gough Whitlam and 'new nationalism'*, Anne Pender, The Australian
Journal of Politics and History, March 2005

'It's very seldom that someone like Barry Humphries comes along' John
Clarke, *Fry's Planet Word*, BBC 2011

'The American influence is huge' Kathy Lette, *Fry's Planet Word*,
BBC 2011

'the practice of spending the night on other people's couches' Vicki Estes,
'Seen the dictionary lately? OMG!' *Topeka Journal*, 10 April 2011

'a boring or socially inept person'; 'the wives and girlfriends' 'Oxford English Dictionary: other new words and definitions', *Telegraph*, 24 March 2011

'*Wag* is notable' Graeme Diamond, editor of Oxford English Dictionary

'a protuberance of flesh' Dave Masters, 'OMG in the OED? LOL!', *Sun*, 24 March 2011

'has apparently taken over from Shakespeare' Mark Liberman, *Homeric Objects of Desire*, 2005

'D'oh ... expressing frustrations' *Oxford English Dictionary*, Oxford University Press, 2001

'The Canadian election was so *meh*' *Collins English Dictionary*, ninth edition, Collins, 4 June 2007

'widespread unselfconscious usage' Graeme Diamond, quoted in 'Meh– the word that's sweeping the internet', Michael Hann, *Guardian*, 5 March 2007

'He's embiggened that role with his cromulent performance' *The Simpsons: Lisa the Iconoclast*, season seven, episode 16, David X Cohen, Fox, 1996

'there is a competing effect' Riccardo Argurio, Matteo Bertolini, Sebastián Franco, Shamit Kachru, *Gauge/gravity duality and meta-stable dynamical supersymmetry breaking*, the Institute of Physics Publishing, 2007

'O hart tht sorz' Eileen Bridge, runner-up in the T - Mobile txt laureate competition, quoted in David Crystal, '2b or not 2b', *Guardian*, 5 July 2008

'Verona was de turf of de feuding Montagues and de Capulet families' Martin Baum, *To Be Or Not To Be, Innit*, Bright Pen, 2008

'John's girlfriend is really pretty' 'Say what? A parents' guide to UK teenage slang', BBC News school report, 11 March 2010

'John's chick is proper buff' Phoenix High School, Shepherd's Bush, West London, 'Say what? A parents' guide to UK teenage slang', BBC News school report, 11 March 2010

'Jonny's bird is proper fit' Holy Family Catholic Church, Keighley, West Yorkshire, 'Say what? A parents' guide to UK teenage slang', BBC News school report, 11 March 2010

'John's missus is flat out bangin'' Bishopston Comprehensive School, Swansea, Wales, 'Say what? A parents' guide to UK teenage slang', BBC News school report, 11 March 2010

'Berkeley is one of the most diverse places you'll ever be' Connor, *Fry's Planet Word*, BBC 2011

'It can change in one day' Berkeley High Students, *Fry's Planet Word*, BBC 2011

'The word rap was used in the black community' H. Samy Alim, *Fry's Planet Word*, BBC 2011

'It's metaphors' Kenard 'K2' Karter, *Fry's Planet Word*, BBC 2011

'The MCing or rapping is a verbal art form' H. Samy Alim, *Fry's Planet Word*, BBC 2011

'You step up into the cipher' ibid.

'From a language perspective' Kenard 'K2' Karter, *Fry's Planet Word*, BBC 2011

'And that is what some people view as cultural theft' H. Samy Alim, *Fry's Planet Word*, BBC 2011

'My president, your country is dead' El General, '*Rais Le Bled*' (President, Your People) quoted in Vivienne Walt, 'El Général and the Rap Anthem of the Mideast Revolution', *Time*, 5 February 2011

'The revolution is a social movement' Balti as quoted by Neil Curry in 'Tunisia's rappers provide a soundtrack to a revolution', CNN World, 2 March 2011

'Egypt, Algeria, Libya, Morocco, all must be liberated' El General, *An Ode to Arab Revolution,* as quoted in *Hip-hop for revolution* by Clark Boyd, PRI's The World, 8 February 2011

Chapter 4

'Without words, without writing and without books, there would be no history' Herman Hesse

'Writing put agreements, laws, commandments on record' H. G. Wells, *A Short History of the World*, Penguin Classics, 2006

'For example, initially there would be a sign' Irving Finkel, *Fry's Planet Word*, BBC 2011

'They were afraid of disease and impotence' ibid.

'Let us ask ourselves, positively, flatly' René Etiemble as quoted in *Writing: The Story of Alphabets and Scripts* by Georges Jean, Harry N. Abrams, March 1992

'Ventris was able to see' John Chadwick, *The Decipherment of Linear B*, Cambridge University Press, 1958

'He who first shortened the labor of copyists' Thomas Carlyle, *Sartor Resartus,* Oxford Paperbacks, June 2008

'for there is so great diversity in English' Geoffrey Chaucer, *Troilus and Criseyde,* Penguin Classics, 2004

'There is an ebb and flow of all conditions of men' Alain-René Lesage, La Valise Trouvée, Imprimerie Nationale, 30 October 2002

'Diderot knew English' Kate Tunstall, *Fry's Planet Word*, BBC 2011

'We know he had these dinners' ibid.

'An encyclopedia' Denis Diderot, *Encyclopedia: the complete illustrations, 1762–1777*, New York: Harry N. Abrams; 1st edition, 1978

'It has been compared to the impious Babel' C.A. Sainte-Beuve, *Portraits of the Eighteenth Century: Historic and Literary, Part II*, translated by Katharine P. Wormeley, G. P. Putnam's Sons, 1905, pp. 89–128

'This is the proportion . . .' Samuel Johnson, quoted in James Boswell, *The Life of Samuel Johnson*, abridged edition, Penguin Classics 31 May 1979

'Wherever I turned my view' Samuel Johnson (1755), Preface to the English Dictionary, paras 1–50, *Dr Johnson's Dictionary*, Penguin Classics, 2005

'. . . to refine our language to grammatical purity' Samuel Johnson, *The Rambler*, No. 208, 14 March 1752

'To make dictionaries is dull work'; 'lexicographer' Samuel Johnson, *Dr Johnson's Dictionary*, Penguin Classics, 2005

'I have protracted my work' Samuel Johnson (1755), Preface to the English Dictionary, *Dr Johnson's Dictionary*, Penguin Classics, 2005

'Dotard'; 'embryo'; 'envy'; 'eavesdropper'; 'or jogger'; 'oats'; 'excise' Samuel Johnson, *Dr Johnson's Dictionary*, Penguin Classics, 2005

'a wretched etymologist' Thomas Macaulay, quoted by Jesse Sheidlower in *Defining Moment*, Bookforum, October 2005

'most truly contemptible performances' John Horne Took, quoted by Henry Hitchings, *Dr Johnson's Dictionary: The Extraordinary Story of the Book that Defined the World*. John Murray, London, 2005

'to the English speaking and English reading public'; 'read books and make extracts for The Philological Society's New English Dictionary' 'Dr Murray, Mill Hill, Middlesex, N.W.' April 1879 Appeal, OED.com

'a1548 Hall Chron., Hen. IV. (1550) 32b, Duryng whiche sickenes as Auctors write he caused his crowne to be set on the pillowe at his beddes heade' William Chester Minor, OED.com

'It's constantly moving. It's fascinating'; 'It's rather nice' John Simpson, *Fry's Planet Word*, BBC 2011

'There's an excitement about the cries and whispers and the solaces' Stephen Fry, *Fry's Planet Word*, 2011

'Books are not absolutely dead things' John Milton, *Areopagitica*, Standard Publications, Inc., July 2008

'was the first to have put together a collection of books' Strabo, quoted by Barbara Krasner-Khait in 'Survivor: The History of the Library', *History* magazine, October/November 2001

'During Lent' Barbara Krasner-Khait, 'Survivor: The History of the Library', *History* magazine, October/November 2001

'I hereby undertake not to remove from the Library' Bodleian Library oath.

'some of those books so taken out by the Reformers were burnt' Anthony Wood, as quoted in 'The History of The Bodleian', bodleian.ox.ac.uk

'set up my Staffe at the Librarie dore in Oxon' Thomas Bodley, as quoted by Jane Curran in 'Looking back on Sir Thomas Bodley', BBC Oxford, June 2009

'We have staff whose job it is to keep stuff safe' Richard Ovenden, *Fry's Planet Word*, BBC 2011

'The politicians of today and tomorrow are communicating with their friends' Richard Ovenden, *Fry's Planet Word*, BBC 2011

'It's always been part of what we see as "serving the whole republic of the learned"' Richard Ovenden, *Fry's Planet Word*, BBC 2011

'Only he who has longed as I did for Saturdays to come' Andrew Carnegie, quoted in *The New York Times* obituary, 12 August 1919

'I choose free libraries as the best agencies for improving the' Andrew Carnegie, as quoted in 'The Andrew Carnegie Story', carnegiebirthplace. com

'Make 'em laugh, make 'em cry, make 'em wait' Wilkie Collins as quoted by GW Dahlquist in 'Make 'em laugh, make 'em cry, make 'em wait', *Guardian*, 6 January 2007

'Is Little Nell dead?' Robert M.C. Jeffrey, *Discovering Tong: Its History, Myths and Curiosities*, Robert Jeffrey, April 2007

'I'm a great goose to have given way so, but I couldn't help it' Lord Jeffrey, as quoted by Hattie Tyng Griswold, *The Lives of Great Authors*, A.C. McClurg and Company, Chicago, 1902

'My taste is for the sensational novel' G. K. Chesterton, *The Spice of Life and Other Essays*, Darwen Finlayson, 1964

'I am an author' Jakucho Setouchi, quoted by Dana Goodyear, 'Letter from Japan, "I ♥ Novels"', *The New Yorker*, 22 December 2008

'The book revolution which from the Renaissance on' John Updike, BookExpo America, 2006

'Print is where words go to die' Jeff Jarvis, 'Books will disappear. Print is where words go to die', *Guardian*, 5 June 2006

'One thing we've learned in the history of books' Professor Robert Darnton, *Fry's Planet Word*, BBC 2011

'I'm very attached to books and to manuscripts' ibid.

'bring back that real book smell you miss so much' Smellofbooks.com

'So we're taking in history through the ear as well as through the eye' Professor Robert Darnton, *Fry's Planet Word*, BBC 2011

'atomic priesthood' Thomas A. Sebeok, 'Technical Report' prepared for Office of Nuclear Waste Isolation Battelle Memorial Institute, April 1984

'People change, culture changes' John Lomberg, *Fry's Planet Word*, BBC 2011

'The other thing that seems to be universal is the notion of a storyboard'
John Lomberg, *Fry's Planet Word*, BBC 2011

Chapter 5

'True Wit is Nature to advantage dress'd,' Alexander Pope, *An Essay on Criticism*, Yale University Press, 1961

'What, old dad dead?' Cyril Tourneur, *The Revenger's Tragedy*, Nick Hern Books, 1996

'Oh that this too too solid flesh would melt, thaw and resolve into a dew!' William Shakespeare, *Hamlet*, 1:2, 129–130, Penguin Classics, 2007

'One might have expected natural selection' Professor Steven Pinker, *Toward a Consilient Study of Literature*, *Philosophy and Literature*, volume 31, Number 1, The Johns Hopkins University Press, 2007

'When you talk about it, you think about it in the back of your head' Ernie Dingo, *Fry's Planet Word*, BBC 2011

'In the time before time began Bunjil' 'The Gariwerd Creation Story' pamphlet

'It's both a rhyme and rhythm, and the rhythm is the heartbeat' Ernie Dingo, *Fry's Planet Word*, BBC 2011

'Tell me, Muse, of the man of many ways, who was driven' Richmond Lattimore, *The Odyssey of Homer*, Harper Collins, 1965, reissued by HarperPerennial 1991, book I, lines 1–10, p. 27

'So they sang, in sweet utterance, and the heart within me desired to listen' ibid., book XII, lines 192–3, p. 190

'Sing, Goddess, Achilles' rage' Professor Stanley Lombardo, *The Iliad*, Hackett Publishing Company, 1997, book I, lines 1–6

'in which the whole plot is done backwards and the story winds up in futility and unhappiness' Professor William Foster-Harris, *The Basic Patterns of Plot*, University of Oklahoma Press, new edition,1981

'Plots of the body'; 'plots of the mind' Ronald Tobias, *20 Master Plots and How to Build Them*, Walking Stick Press, March 2003

'Nobody knows anything' William Goldman, *Adventures in the Screen Trade*, Futura Publications, 1990

'Nobody – not now, not ever – knows the least goddamn thing about what is or isn't going to work at the box office' ibid.

'Writing is finally about one thing: going into a room alone and doing it' William Goldman, *Fry's Planet Word*, BBC 2011

'I was walking on 47th Street in New York – the diamond district' ibid.

'Is it safe?' William Goldman, *Marathon Man*, Paramount Pictures, 1976

'There's no logic to it' William Goldman, *Fry's Planet Word*, BBC 2011 'disappeared off with the Indians' ibid.

'But who knows, if we'd had McQueen, if it would have been different'
ibid.

'When I tried to sell the movie' ibid.

'I'm gonna say something stupid' ibid.

'just to let themselves go and swim into it' ibid.

'I didn't know Joyce, I didn't know his wife Nora' David Norris, *Fry's
Planet Word*, BBC 2011

'Few writers have had more grace and splendour in the way they write'
ibid.

'"Good day's work, Joyce?" said Budgen' ibid.

'Every kind of Dublin saying, like "the sock whiskey" for sore legs, for
instance, is in it' ibid.

'Mr Leopold Bloom ate with relish' James Joyce, *Ulysses*, Penguin
Classics, new edition, 2000

'My soul frets in the shadow of his language' James Joyce, *A Portrait of
the Artist as a Young Man*, Wordsworth Editions Ltd, new edition, May
1992

'I have no language now, Sheila' David Norris, *Fry's Planet Word*,
BBC 2011

'We tend to be a bit subversive' ibid.

'I would say the greatness of Yeats' Declan Kiberd, *Fry's Planet Word*,
BBC 2011

'And for me that's the magnificence of Yeats' ibid.

'We wouldn't be here today in a Senate of an independent Ireland' ibid.

'I have met them at close of day' W.B. Yeats, *Easter, 1916 – WB Yeats
Selected Poems*, Penguin Classics, reissued 2000

'Sure, but you have to tell us a story in return' Declan Kiberd, *Fry's Planet
Word*, BBC 2011

'You make your destitution sumptuous' ibid.

'In the last ditch, all we can do is sing' Samuel Becket as quoted by Barry
McGovern, *Fry's Planet Word*, BBC 2011

'Absolutely terrifying ... Apart from it being so famous' Simon Russell
Beale, *Fry's Planet Word*, BBC 2011

'I get the sense that it was a radical exploration of a single human soul'
ibid.

'It really does yield extraordinary riches' ibid.

'Utterly terrifying, poleaxing with fear' David Tennant, *Fry's Planet Word*,
BBC 2011

'I saw him at that very formative age' ibid.

'like keeping goal for Scotland' ibid.

'And you think, please Lord, let me just remember the lines' ibid.

'I can be bounded in a nutshell' William Shakespeare, *Hamlet*, 2:2,
254–6, Penguin Classics, new edition, January 2007
'You just get the sense that he hasn't slept for days' David Tennant, *Fry's
Planet Word*, BBC 2011
'honey-heavy dew of slumber' William Shakespeare, *Julius Caesar*, 1:1,
230, Wordsworth Editions, 1992
'sore labour's bath' William Shakespeare, *Macbeth*, 2:2, 35–6, Penguin
Classics, new edition, 2007
'Alas poor Yorick, I knew him' William Shakespeare, *Hamlet*, 5:1, 185,
Penguin Classics, new edition, 2007
'I think the Yorick moment is much more specific' David Tennant, *Fry's
Planet Word*, BBC 2011
'The first few performances holding a real human head' ibid.
'like it was some big oak tree' Mark Rylance, *Fry's Planet Word*, BBC 2011
'"It's you who are alive now."' ibid.
'In Pittsburgh there was a little old lady' ibid.
'Till then, sit still, my soul: foul deeds will rise' William Shakespeare,
Hamlet 1:2, 256–7, Penguin Classics, new edition, January 2007
'I remember performing the play out at Broadmoor special hospital'
Mark Rylance, *Fry's Planet Word*, BBC 2011
'To be, or not to be, that is the question' William Shakespeare, *Hamlet*,
3:1, 56–88 Penguin Classics, new edition, January 2007
'I think poetry is to be distinguished always from prose.' Christopher
Ricks and Stephen Fry, *Fry's Planet Word*, BBC 2011
'Stop all the clocks, cut off the telephone' W. H. Auden, 'Funeral Blues',
Collected Poems, Faber & Faber, Copyright © 1976, 1991, 2007 The
Estate of W. H. Auden
'Tragically in my life, in every film I've ever done' Richard Curtis, *Fry's
Planet Word*, BBC 2011
'Every day I think of that line from "The Boxer"' ibid.
'If you pick up a poem for the first time you have to piece it together' ibid.
'I think there was a six-month period in which I understood it' ibid.
'music is auditory cheesecake' Professor Steven Pinker, *How the Mind
Works*, Penguin Allen Lane, 1998
'A Mars a day helps you work, rest and play' Mars, D'Arcy Masius
Benton & Bowles, 1965
'Now hands that do dishes . . .with mild green Fairy Liquid' Fairy Liquid
advertising campaign, 1961
'Just do it' Nike, Wieden and Kennedy, 1988
'Most excellent and proved Dentifrice' *London Gazette*, 1660, quoted
by Gillian Dyer, *Advertising as Communication*, Routledge, new
edition, 1982

'Promise, large Promise, is the soul of an Advertisement' Dr Samuel
   Johnson, *The Idler,* The Universals Chronicle, 1759 edition
'It won't be long till Leo Burnett is selling apples' quoted in *The Apple
   Story,* leoburnett.co.in
'They're GR-R-R-E-A-T' Kellogg's Frosties advertising campaign, Leo
   Burnett, 1952
'Words are tremendously important in advertising' Don Bowen, *Fry's
   Planet Word,* BBC 2011
'We pluck the lemon; you get the plums' Volkswagen advertising
   campaign, Doyle Dane Bernbach, 1960
'We're only Number 2. We try harder' Avis advertising campaign, Doyle
   Dane Bernbach, 1962
'Probably the best lager in the world' Carlsberg advertising campaign,
   Saatchi & Saatchi, 1973
'Refreshes the parts other beers cannot reach' Heineken advertising
   campaign, Collett Dickenson Pearce & Partners, 1974
'Don't just book it, Thomas Cook it' Thomas Cook advertising
   campaign, Wells, Rich, Greene, 1984
'English is a particularly good language for being able to play gags' Don
   Bowen, *Fry's Planet Word,* BBC 2011
'Don't forget the Fruit Gums, Mum' Rowntree's Fruit Gums advertising
   campaign, 1958
'Beanz Means Heinz' Heinz advertising campaign, Young & Rubicam,
   1967
'All because the lady loves Milk Tray' Cadbury's advertising campaign,
   Leo Burnett, 1968
'Opal Fruits! Made to make your mouth water' Mars advertising
   campaign, *c.*1970
'For Mash get Smash!' Smash advertising campaign, Boase Massimi
   Pollitt, 1974
'Happiness is a cigar called Hamlet' Hamlet advertising campaign, Collett
   Dickenson Pearce & Partners, 1960
'The end line has got to resonate with people' Don Bowen, *Fry's Planet
   Word,* BBC 2011
'Now the laborers and cablers and council-motion tablers were just
   passing by' McDonald's, Leo Burnett, 2009
'I think there will always be good stories to tell' Don Bowen, *Fry's Planet
   Word,* BBC 2011
'Unless your advertising contains a big idea it will pass like a ship in the
   night' David Ogilvy, *Ogilvy on Advertising,* Prion Books, 1995
'In Great Britain, there are twelve million households' David Ogilvy, *The
   Theory and Practice of Selling the Aga Cooker,* issued by Aga Heat
   Limited, June 1935

'no credentials, no clients and only $6,000 in the bank' David Ogilvy, *An Autobiography*, John Wiley & Sons; Revised Ed edition, February 1997

'Did it make me gasp when I first saw it?' David Ogilvy, *Ogilvy on Advertising*, Prion Books, 1995

'The Man in the Hathaway shirt' C. F. Hathaway, Ogilvy and Mather, 1951

'At sixty miles an hour the loudest noise in this new Rolls-Royce comes from the electric clock' Rolls Royce advertising campaign, Ogilvy & Mather, 1958

'Only Dove is one-quarter cleansing cream' Dove advertising campaign, Ogilvy & Mather,

'The consumer is not a moron' David Ogilvy, *Ogilvy on Advertising*, Prion Books, 1995

'Being edited by Ogilvy was like being operated on by a great surgeon' Kenneth Roman, *The King of Madison Avenue: David Ogilvy and the Making of Modern Advertising*, Palgrave Macmillan, 2009

'I do not regard advertising as entertainment or art form' David Ogilvy, *Ogilvy on Advertising*, Prion Books, 1995

'David Ogilvy 1911 – Great brands live for ever' Leo Burnett, 1999

'A lot of communication has nothing to do with the words' President Clinton, *The Art of Oratory*, BBC News, April 2009

'Friends, Romans, countrymen, lend me your ears!' William Shakespeare, *Julius Caesar*, 3:2, 73, Wordsworth Editions Ltd, 1992

'To be or not to be' *Hamlet*, 3:1, 56, Penguin Classics, 2007

'Out, damn spot' William Shakespeare, *Macbeth*, 5:1, 35, Penguin Classics, 2007

'Brutus is an honourable man' William Shakespeare, *Julius Caesar*, 3:2, 87, Wordsworth Editions Ltd, 1992

'Words paint pictures; words draw our imagination' The Reverend Jesse Jackson as quoted in *The Art of Oratory*, BBC News, April 2009

'It is this fate, I solemnly assure you, that I dread for you' Demosthenes, translated by Arthur Wallace Pickard in *The Public Orations of Demosthenes, Volume I*, The Echo Library, January 2008

'Four score and seven years ago our fathers brought forth on this' Abraham Lincoln, *The Gettysburg Address*, Penguin, August 2009

'I am tired of fighting. Our Chiefs are killed; Looking Glass is dead' Kent Nerburn, *Chief Joseph & the Flight of the Nez Perce: The Untold Story of an American Tragedy*, HarperOne, 2006

'We shall not flag or fail' Winston Churchill in the House of Commons, 4 June 1940, as quoted in David Cannadine (ed.), *Blood, Toil, Tears and Sweat: The Great Speeches*, Penguin Classics, 2007

'He mobilized the English language and sent it into battle' Edward R. Murrow, CBS broadcast, 30 November 1954, quoted in David Cannadine (ed.), *Blood, Toil, Tears and Sweat: The Great Speeches*, Penguin Classics, 2007

'that branch of the art of lying which consists in very nearly deceiving your friends without quite deceiving your enemies' Francis Cornford, quoted by Michael Balfour in *Propaganda in War, 1939–45: Organizations, Policies and Publics in Britain and Germany*, Routledge, 1979

'You know what "morale" is, don't you?' Graham Greene, John Dighton, Angus MacPhail, Diana Morgan, *Went the Day Well?*, Ealing Studios, 1942

'It's a beautiful thing, the destruction of words' George Orwell, *1984*, Penguin, 2008

'In the end we shall make thoughtcrime literally impossible' ibid.

'partly to improve his French'; 'and it was an ideology, not just a language' Bernard Crick, *George Orwell: A Life*, Penguin, 2nd revised edition, 1992

'Who controls the past controls the future; who controls the present controls the past' George Orwell, *1984*, Penguin, 2008

'That propaganda is good which leads to success' Joseph Goebbels, quoted by Joachim C. Fest, *The Face of the Third Reich*, Penguin, new edition, 1995

'Letting a hundred flowers blossom' Roderick MacFarquhar, *The Hundred Flowers Campaign and the Chinese Intellectuals*, Praeger, 1960

'War is peace. Freedom is slavery. Ignorance is strength' George Orwell, *1984*, Penguin, 2008

'Political slogans are still useful because they sum up the approach' Don Bowen, *Fry's Planet Word*, BBC 2011

'The most brilliant propagandist technique' Adolph Hitler, *Mein Kampf*, Jaico Publishing House, 37th edition, 2007

Bibliography

Aldridge, David (ed.), *Music Therapy in Dementia Care*, Jessica Kingsley 2000

Allen, Keith and Burridge, Kate, *Forbidden Words: Taboo and the Censoring of Language*, Cambridge University Press 2006

Auden, W. H., *Collected Poems*, Faber and Faber 1976

Barrett, Anthony, *Turkana Iconography*, A.J. Barrett 1997

Baum, Martin, *To Be or Not To Be, Innit*, Bright Pen 2008

Bauer, Laurie and Trudgill, Peter (eds), *Language Myths*, Penguin 1998

Berkley High School Slang Dictionary, North Atlantic Books 2004

Boyle, T. C., *Wild Child*, Bloomsbury 2010

Brewer's Dictionary of Phrase and Fable (18th edition), Chambers Harrap 2009

Bryson, Bill, *Mother Tongue*, Penguin 1990

Burling, Roberts, *The Talking Ape*, Oxford University Press 2005

Burton, Peter, *Parallel Lives*, Gay Men's Press 1985

Cavalli-Sforza, Luigi Luca, *Genes, Peoples and Languages*, Penguin 2000

Chomsky, Noam, *Architecture of Language*, Oxford University Press 1996

Crystal, David, *How Language Works*, Penguin 2006

——, *Txting*, Oxford University Press 2008

——, *The Cambridge Encyclopedia of Language* (3rd edition), Cambridge University Press 2010

——, *The Fight for English*, Oxford University Press 2006

Deutscher, Guy, *Through the Language Glass*, William Heinemann 2010

——, *The Unfolding of Language*, William Heinemann 2005

Eco, Umberto, *The Search for the Perfect Language*, Blackwell 1995

Fagles, Robert (trans.); introduction by Bernard Knox, *The Odyssey*, Viking Penguin 1996

Fitch, W. Tecumseh, *The Evolution of Language*, Cambridge University Press 2010

Forster, E. M., *Aspects of the Novel*, Penguin Classics 2005

Frommer, Paul and Finegan, Edward, *Looking at Languages* (4th edition), Thomson/Wadsworth 2008

Goldman, William, *Adventures in the Screen Trade*, Grand Central Publishing 1983

Gower, Ernest, *The Complete Plain Words* (3rd revised edition), Penguin 2004

Gregory, Richard (ed.), *The Oxford Companion to the Mind*, Oxford University Press 1987

Grose, Francis, *A Classical Dictionary of the Vulgar Tongue*, printed for Hooper and Wigstead, 1796 Google ebook

Harris, Roy, *Rethinking Words*, The Athlone Press 2000

Hitchins, Henry, *The Language Wars*, John Murray 2011

——, *The Secret Life of Words*, John Murray 2008

——, *Dr Johnson's Dictionary*, John Murray 2005

Hughes, Geoffrey, *Swearing*, Penguin 1991

Jacot de Boinard, Adam, *I Never Knew There Was a Word For It*, Penguin 2005

Jay, Timothy, *Why We Curse*, John Benjamins Publishing 2000

——, *Cursing in America*, John Benjamins Publishing 1992

Jay-Z, *Decoded*, Random House 2010

Jean, Georges, *Writing – The Story of Alphabets and Scripts*, Thames and Hudson 1997

Joyce, James, *Ulysses*, Wordsworth Editions Limited 1932

——, *A Portrait of the Artist as a Young Man*, Penguin Modern Classics, 2000

Keyes, Ralph, *Euphemania*, Little, Brown and Co. 2010

Klamperer, Victor, *Lingua Tertii Imperii (The Language of the Third Reich)*, Continuum International 2002

Kipling, Rudyard, *Just So Stories*, Wordsworth Editions 1993

Korzhenov, Aleksander Zamenhof, *The Life, Works and Ideas of the Author of Esperanto*, Mondial 2010

Kurlansky, Mark, *The Basque History of the World*, Jonathan Cape 1999

Lattimore, Richard, *The Odyssey of Homer*, HarperCollins 1965

Leach, Edmund, *Culture and Communication*, Cambridge University Press 1976

Levitin, Daniel, *This is Your Brain on Music*, Dutton 2006

Mitford, Nancy (ed.), *Noblesse Oblige: An Enquiry Into The Identifiable Characteristics of the English Aristocracy*, Hamish Hamilton 1956

Norrière, Jean-Paul, *Parlez Globish* (2nd edition), Groupe Eyrolles 2006

——, and Hon, David *Globish The World Over*, International Globish Institute, 2009

Newton, Michael Savage, *Girls and Wild Boys*, Faber and Faber 2002

Ogilvy, David, *Ogilvy on Advertising* (new edition), Prion Books Ltd 1995

Okrent, Arika, *In the Land of Invented Languages*, Spiegel and Grau 2009

Orwell, George, 'Politics and the English Language', in *Essays*, Penguin Modern Classics, 2000

——, *1984*, Penguin Modern Classics, new edition, 2004

Ostler, Nicholas, *Empires of the Word*, HarperCollins 2005

Pickles, Wilfred, *Between You and Me*, Werner Laurie, 1949

Pinker, Steven, *How the Mind Works*, Penguin 1997

——, *The Language Instinct*, Penguin 1994

——, *The Blank Slate*, Penguin 2002

——, *The Stuff of Thought*, Penguin 2007

Ramachandran, V. S. and Blakeslee, Sandra, *Phantoms in the Brain*, Fourth Estate 1998

Ricks, Christopher, *The New Oxford Book of Victorian Verse*, Oxford University Press 1987

——, *Reviewery*, Penguin 2003

——, *Dylan's Vision of Sin*, Collins 2004

Roman, Kenneth, *The King of Madison Avenue*, Palgrave Macmillan 2009

Rossiter, A. P., *Our Living Language*, Longman's, Green and Co. 1953

Sandler, Wendy and Lillo-Martin, Diane, *Sign Language and Linguistic Universals*, Cambridge University Press 2006

Schaller, Susan, *A Man Without Words*, Simon and Schuster 1990

Sebeok, Thomas, *Speaking of Apes*, Plenum Press 1980

Shattuck, Roger, *The Forbidden Experiment*, Farrer, Straus and Giroux 1980

Silverton, Peter, *Filthy English*, Portobello Books 2009

Singh, Simon, *The Code Book*, HarperCollins 1999

Steiner, George, *After Babel*, Oxford University Press 1998

Thorne, Tony, *Dictionary of Contemporary Slang* (3rd edition), A & C Black Publishers 2009

Tomasello, Michael, *Origins of Human Communication*, MIT Press 2008

Wells, H. G., *A Short History of the World*, Penguin Classics 2006

Werner, Stephen, *Blueprint: A Study of Diderot and the Encyclopedie Plates*, Summa Publications, Inc. 1993

Whorf, Benjamin, *Language, Thought and Reality*, MIT Press 1964

Winchester, Simon, *The Professor and the Madman*, HarperCollins 1998

Index

Page references in **bold** indicate illustrations.

Acknowledgements

Planet Word is a companion work to the BBC series *Fry's Planet Word*. Without Stephen Fry, none of it would have happened. His passion, erudition, wit and curiosity inform every page of the book. Thank you, Stephen. My special appreciation goes to Margaret Magnusson and Anna Magnusson, who helped beyond measure in making sure the words were put down on paper. Laura Herring and Louise Moore at Penguin shaped and edited the book, and Anthony Goff smoothed the whole process.

The book, like the television series, is a collective enterprise. Helen Williamson, my co-director, and the production team of Annie Macnee, Lucy Wallace, Lucy Tate and Clare Bennett all worked tirelessly and joyfully making the films happen. Simon Ffrench and Adam Toy made the sound and pictures sing, whilst Masahiro Hirakubo, Mikhael Junod and Paul Burgess skilfully edited the hours and hours of material.

At Sprout Pictures, Gina Carter, Zoe Rocha and Emily Martin provided back-up whenever it was needed. I'm particularly grateful to Janice Hadlow and Mark Bell at the BBC for having the courage to commission a series about words.

So many people, too numerous to mention individually, have contributed their time and expertise over the past two years; they are the heart and soul of the series and the book. Arika Okrent, Steven Pinker, Declan Kiberd, Paul Frommer, Ray Dolland and Cathy Price were ever generous with their time. In particular I would like to thank Nigel Williams, Robert McCrum, Guy Deutscher and Sir Christopher Ricks for their help and insights throughout.

And lastly, thank you, Robbie, Calum, Ellie and Louis, who have had to bear so many conversations *about* language that my wife, Margaret Magnusson, and I often forgot to talk *to* them.